Dear Nana Wendy
Happy Birthday Nov
love you lots
Callum ⊗⊗⊗

Australia's
VANISHING MAMMALS

Australia's VANISHING MAMMALS

ENDANGERED AND EXTINCT NATIVE SPECIES

Dr Tim Flannery

Paula Kendall

Karen Wynn-Moylan

RD PRESS

First published in Australia in 1990 by RD Press
a registered business name of
Readers Digest (Australia) Pty Limited (Inc. in NSW)
26-32 Waterloo Street, Surry Hills, NSW 2010

ISBN 0 86438 161 1

This book was designed and produced by Bow Press
208 Victoria Road, Drummoyne

National Library of Australia Cataloguing-in-Publication Data

Flannery, Timothy Fridtjof.
Australia's vanishing mammals.

Includes index.
ISBN 0 86438 161 1.

1. Rare mammals — Australia. 2. Extinct mammals — Australia.
I. Kendall, Paula. II. Wynn-Moylan, Karen, 1953- . III. Title.

599.00420994

PREFACE

Not all the species that feature in *Australia's Vanishing Mammals* were known at the time John Gould (1804-1881) was researching Australian mammals. Of the 30 species in this book, 22 were illustrated in *Mammals of Australia,* by J. Gould and H. C. Richter, published 1840-1863 in 13 parts. These hand coloured lithographs, although beautiful, were not always scientifically accurate. They have been included in *Australia's Vanishing Mammals* for historical interest.

The degree to which a species is at risk of becoming extinct cannot be measured solely in numbers of individuals surviving. This is because some animals, like the Tasmanian devil, would naturally have a more sparse population than others, such as the rufous hare-wallaby, yet due to its history and biology, the wallaby is considerably more endangered.

Anyone who thinks they recognise an endangered or extinct species, or an animal not previously recorded in a particular habitat should contact the closest appropriate institution.

FOREWORD

JIM FLANNERY is charitable to the human race in his choice of title for this book; he might well have described Australia's many now extinct species of mammals as having been actively destroyed rather than as having merely "vanished". The record of the past, told in these pages in terms of mammals, is an evil one. But perhaps humanity deserves some charity, to the extent at least that we are coming to see the evil of our ways, and are changing them.

There are, I suppose, at least two ways of viewing ecodiversity and the preservation of species. One way is through eyes that are focused exclusively upon mankind, upon our material needs and our aesthetic pleasures, both so dependent upon the living ecosystems all around us. The other is to look at ecodiversity as no more concerned with mankind than with any other living thing on Earth, and to treat the preservation of species in all their extraordinary diversity as a sufficient end in itself, regardless of the needs of mankind, just as one might care for Earth itself for its own sake and not simply as the only habitat for living things in our solar system.

While philosophically poles apart, either of these views must surely point in the same direction, encouraging every thinking member of the human race to act now to stop any further wholesale, man-made destruction of species. In this last decade of the twentieth century we not only possess the knowledge, the social organisation and the technical ability to bring this continuing destruction to an end; we are also blessed by a rising tide of awareness among people world-wide that we must do just that — that we must prevent the destruction of further species and must care for presently endangered species.

It is this new awareness of the problem that has now to be followed up by evermore vigorous action, both domestically and on the international scene. An international convention on biodiversity is already in the making, and in the Australian community there are already many working actively to ensure that our remaining species of fauna and flora, many of them unique to our continent, are protected.

If this book further sharpens existing awareness of the importance of biodiversity, gives further encouragement to those already working in the field and inspires others to join them, Tim Flannery will have well served both the human race and the bountiful Earth that mankind has too long taken for granted.

Rt. Hon. Sir Ninian Stephen

CONTENTS

INTRODUCTION

THE FIGURES are simple but shocking: of the 60 mammal species that have become extinct worldwide in the last 500 years, 20 were Australian. Indeed, in the field of animal conservation, Australia's record is deplorable. No other country has so rapidly put so much of its wildlife under threat.

Until recently, this decline was accepted as an inevitable cost of "progress", an economic necessity in the process of creating jobs and developing the nation. Now, after 200 years of intensive grazing, forestry and mining, with little regard for their effects on our wildlife, fundamental attitudes are at last changing. We — and nations worldwide — are beginning to understand just what the extinction of our animals represents.

Many people argue that the vanished species, although interesting and attractive, were useless, and that the loss of a few rats and wallabies has no effect upon humans. Nothing could be further from the truth. We are learning to our cost that the human animal cannot exist in isolation. Destroy the environment that supports other creatures and we put ourselves under threat. Degrade the land and we cannot use it to grow food; poison the rivers and seas and we lose our fish and seabirds; cut down the trees and we increase air pollution and we even change our climate. Ecosystems destabilised by extinction are not robust: they can continue to degrade with even minor infringement. Vanishing wildlife is a first and obvious sign of a serious environmental malaise, one that will deepen as extinctions continue, so that eventually human existence itself is placed in jeopardy.

In this book you will find detailed information on a representative selection of extinct and endangered native mammals. The book is not just a cavalcade of curious creatures — nor is it intended to be. Its purpose is to increase public awareness of the perilous state of many species and to highlight the need for proper conservation practices.

Only time will show the full effects of mammal extinctions on this continent. In vast parts of central Australia, there are now virtually no living native mammals between the size of a rat and a red kangaroo. The lost species filled important places in the food chain, a chain that ultimately includes humans. For instance, many insect-eating and carnivorous mammals have died out, and some plant species seem to be no longer reproducing successfully, perhaps under the pressure of unbridled insect and mammalian herbivore attack. If this trend continues, the arid and semi-arid zones will become less stable; soil loss may accelerate, bringing eventually a Sahara-like dunefield. The grazing of sheep and cattle — two of Australia's top export earners — will be impossible. And the effects won't stop there. Unfortunately, extinction is forever: the actions and reactions set in train are complex and unpredictable.

Conservation measures can be effective, but we need to know more about how and why our lost species became extinct. We do know that European settlers drastically changed the way Australia's land was used. Aborigines were removed from their lands, forests and ground cover were cleared, exotic grasses were planted, sheep and cattle introduced, animal pets let loose. Wetlands were drained, wilderness regions destroyed. Vast wheat farms were created, agriculture by irrigation was developed, mines were sunk, factories built, cities founded. And yet the details of this broad picture need to be set in sharper focus. What at first glance seems obvious and unquestionable often proves on closer examination to be quite different.

For example, hunting and introduced predators such as the fox and the cat have been blamed for wiping out many of our animals. Recent research indicates, however, that these factors have perhaps had a less important impact, except on the thylacine (Tasmanian tiger), which was probably hunted and trapped to extinction, and the toolache wallaby, whose struggling survivors were shot for sport. Several mammal species became extinct long before the fox crossed the Murray River and invaded most of Australia in the 1890s; other species coexisted for years with feral cats without apparent decline.

If we look in detail at two remote regions we may discover some of the complex reasons behind the threat to our wildlife. In the Tanami and Great Victoria Deserts, stock animals never grazed and foxes and rabbits had minimal impact. The native wildlife survived intact until at least the 1950s, when the traditional Aboriginal owners were persuaded to leave and live in mission stations. When some Aborigines returned in the 1980s, many species had vanished. Scientists believe that their disappearance was caused by uncontrolled fire.

For 40,000 years, Aboriginal people practised firestick farming throughout Australia. Continuously, they lit small and comparatively cool fires, burning the land in a complex mosaic, flushing out game. Surprisingly, we might think, these small fires benefited most of the animals, encouraging their food supplies and ensuring shelter. In turn, the Aborigines were guaranteed a constant food source. More importantly, perhaps, controlled burning meant that large fires were prevented. When Europeans forced Aborigines off their lands, the vegetation built up enormously. Following good seasons, vast wildfires broke out, burning millions of hectares in the arid zone. Although birds and kangaroos migrated, middle-sized mammals could not. Forced to forage for scarce resources in ever more-exposed locations, their numbers diminished through increased vulnerability to predators. Eventually, they became extinct throughout their range.

Many questions remain unanswered. Why, for instance, did some species survive while similar ones perished, and why were there more extinctions in the desert than elsewhere? We need a national research programme to provide the answers. Armed with precise and accurate information, we will be better placed to formulate policies that will stabilise our fragile ecosystems.

We may find it necessary to modify practices that have contributed to the processes that threaten wildlife with extinction. For example, if the raising of sheep and cattle is implicated as a major factor, we may need to change our farming methods — even to farm large kangaroos and emus as alternatives. The move would be controversial, but it has already been shown that we can use these common native species without endangering them. To combat soil erosion and salination, we will need also to change the way we use our land, replanting trees where we have cut them down and retaining them where once they would have been felled. Ultimately, we must make sure that we do not ask too much from our fragile land.

Already, there are encouraging signs of change and we have many bright successes to guide us in the future. Fauna reserves have been established and national parks expanded to include the ranges of threatened species. Captive breeding programmes and management schemes, including the reintroduction of controlled fire in some areas, show great promise. Daily, our wildlife scientists are increasing the knowledge that will enable them to advise governments and other institutions about how to manage our environment.

Not everything is in the hands of experts. More and more, Australians are discovering that they have an important role to play in protecting the land they live in. If you think you recognise an endangered species or one thought to be extinct, contact an appropriate institution, such as the national parks and wildlife service, a natural-history museum or a university department. Tell them also if you see an animal that has not previously been recorded in that habitat. Much valuable information has been gained in this way.

This book presents a depressing picture. The encouraging factor is that we have recognised the crisis in our society and are changing our attitudes to overcome it. That alone holds out hope for our endangered wildlife and, if we succeed, will preserve our own increasingly precarious place in the world.

EXTINCT NATIVE SPECIES

CRESCENT NAIL-TAIL WALLABY

(Onychogalea lunata)

FROM
The Mammals of Australia,
John Gould, 1848-1863.
Gould's name for the mammal:
Lunated Nail-tail Kangaroo,
Onychogalea lunata.

When chased this wallaby would take refuge in hollow logs — an unusual habit for a medium-sized member of the kangaroo family.

Last sighting 1956

Vulnerable	*Could become endangered because few in numbers or limited in distribution.*
Endangered	*Likely to become extinct unless action is taken.*
Critical	*Only a few hundred individuals survive, or only a single small population known.*
Extinct	*Not sighted for more than 50 years.*

The great naturalist John Gould described this species in 1841 from specimens collected by his employee, John Gilbert, in the southwest of Western Australia. Measurements for a mixed sex sample of crescent nail-tail wallabies are: head and body length 371-508 mm, with the tail adding a further 153-330 mm. There appears to have been little difference in size between males and females. No information is available about the weight of this species. This small wallaby was similar in appearance to the bridled nail-tail wallaby, but was less brightly coloured and lacked a black back stripe. In size and appearance this very slight and delicate species was rather similar to a large rabbit, and had soft, fine fur. The overall colour was ashy grey; the underfur was long and slaty grey at the base, grading to pale grey at the tips. There was a pale cheek stripe and a pale area around the eyes, and the relatively short ears were white inside and brown on the back. The back and sides of the neck were a rich reddish colour. Both its common and scientific names were inspired by the prominent, well-defined and crescent-shaped white stripe on the shoulder. There was also a double, but poorly defined, hip-stripe. The relatively small tail was a uniform grey colour with longer and darker hairs forming a slight crest at the end. There was also a horny nail at the very tip of the tail — a feature of all the nail-tail wallabies.

Unfortunately, as a result of its early demise, nothing was recorded of its reproduction. Information on other aspects of its lifestyle is also scant. Early reports indicated that when chased this wallaby would take refuge in hollow logs — an unusual habit for a medium-sized kangaroo. Aborigines in the south-west of Western Australia used to light fires and smoke out the *wurrung*, as they called it. Gilbert also observed another of its unusual habits: while sitting quietly and cleaning itself, the crescent nail-tail wallaby constantly twitched its tail in an upward direction — something that Gilbert, an experienced naturalist, never saw in any other kangaroo species.

Gilbert noted that the crescent nail-tail wallaby 'is found in the gum forests of Western Australia, where there are patches of thick scrub and dense thickets. In the open glades intervening between them it is occasionally seen sunning itself, but at the slightest alarm immediately betakes itself to the shelter of the thick scrub . . .' It may have favoured areas where dense scrub and open woodland occur together, as does its relative, the bridled nail-tail wallaby. Central Australian Aborigines recall that in the more arid part of its range

CRESCENT NAIL-TAIL WALLABY

Onychogalea lunata

Head and body length: 371-508 mm
Tail length: 153-330 mm

Painting dimensions: 382 x 242 mm

Reconstruction: a stuffed skin was examined for details of markings; the anatomy was compiled from observation and photographs of the nearest living relative, the bridled nail-tail wallaby.

The now extinct crescent nail-tail wallaby was once abundant in mallee scrub such as this near Peak Charles in the southwest of Western Australia.

(A. Burbridge & J. Raines/ANT Photo Library)

this wallaby inhabited mulga country and areas near creeks with red river gums. It does not appear to have made nests, but instead may have waited out the heat of the day in a hollow formed in soft ground beneath a bush.

Prior to European settlement the crescent nail-tail wallaby had a wide, although apparently patchy, distribution. The majority of early records are from the southwest of Western Australia and it appears to have been quite plentiful in the agricultural districts there until just after the turn of the century. Twenty-three specimens were collected from this area for the British Museum between 1904 and 1907. However, the species underwent a rapid decline soon after and the last recorded specimen was taken from there in 1908. Gerard Krefft records it from the west of the Darling River in New South Wales during the Blandowski Expedition in 1856-57, and a mounted

specimen, probably Krefft's animal, is now kept in the Museum of Victoria. There are records of it from the Northern Territory and the northern edge of the Nullabor Plain, the latter in 1929. The Horn Expedition collected it in the Alice Springs area in 1894 and Hedley Finlayson records one being killed between the Tarlton and Jervois Ranges in the Northern Territory as late as 1956. This is the last unequivocal record of its existence, although there were spurious reports of fresh remains collected from Western Australia in the 1960s. The Elder Expedition of 1891 collected it in the Everard Ranges in northwestern South Australia while Professor F. W. Jones reports upon three specimens collected by Sir George Grey in South Australia which were sent to the British Museum. In the 1930s Finlayson reported it as still existing in the Everard and Musgrave Ranges in South Australia and the Cavenagh Range in Western Australia. Both Western Australian and Northern Territory Aborigines state that they have not seen it for many years. Jones, writing in 1924, concluded that the species was probably already extinct in South Australia and rapidly disappearing from Western Australia. Unfortunately no one had the foresight to protect the survivors in a sanctuary. As Jones commented: 'I know of no attempt to preserve this Western Australian remnant by placing some of the animals in a faunal sanctuary'.

Although a single sighting has been made in the past 50 years, the species is now considered most likely to be extinct. The reasons for its disapperance remain unclear.

While sitting quietly and cleaning itself, it constantly twitched its tail in an upward direction.

DESERT BANDICOOT

(Perameles eremiana)

The desert bandicoot was once found in areas such as this newly burned spinifex grassland in central Australia. It preferred sandplain and sandridge desert with spinifex grassland.

(K. Johnson/NPIAW)

It was sometimes called the orange-backed bandicoot, because the rump sported an orange patch.

The desert bandicoot, which is also sometimes called the orange-backed bandicoot, was first described scientifically by Professor Baldwin Spencer of Melbourne University in 1897. He distinguished this slenderly built bandicoot from all other species of *Perameles* by the dull orange colour of its upper body. It had soft fur and numerous, long dark, spiny hairs on the top of the head and body that continued as two dark bands down each side of the rump. The margin of the upper jaw and underside of the head and body were white, as were also the inner side of the back legs and the upper surface of the feet. The rump sported an orange patch while the soles of the hind feet were covered with dark brown hairs that extended to the pads at the base of the fourth and fifth toes. The hairy soles of the hind feet may have been adapted for life in an arid, sandy habitat. The first and fifth toes of the front feet were very short and were without claws. The tail was bi-coloured; the upper surface was blackish while the undersurface and sides were white. The muzzle was fairly long and slender and the ears long, narrow and pointed. The head and body measurements recorded for two specimens are 235 mm for males and 275 mm for females. There is no recorded weight for this species, as all known specimens were collected before weighing specimens became a part of zoological routine.

Although the female's pouch contained eight nipples, two is the only recorded litter size for this species. Reports indicate that it built its nests in hollows which were scooped out in the sand and lined with grass and twigs. It appears to have been nocturnal and to have rested in its nest during the day. There are recorded descriptions of Aborigines catching desert bandicoots by placing one foot on the nest to pin the animal down, then pulling it out by hand.

The desert bandicoot's original habitat was sandplain and sandridge desert with spinfex grassland. This habitat extended across an arid belt of central Australia from Alice Springs in the east to the Rawlinson and McIllwraith ranges, Western Australia, in the west. The original specimens came from Burt Plain, north of Alice Springs, and from sandhills about 65 km to the northeast of Charlotte Waters. The full extent of its original range will probably never be known. It appears to have been locally common in some parts of the central desert as late as the 1930s and there are unverified reports of sightings, by the biologist and explorer Hedley Finlayson, from near the Granites

Last sighting 1931

Vulnerable — *Could become endangered because few in numbers or limited in distribution.*

Endangered — *Likely to become extinct unless action is taken.*

Critical — *Only a few hundred individuals survive, or only a single small population known.*

Extinct — *Not sighted for more than 50 years.*

DESERT BANDICOOT

Perameles eremiana

Head and body length: 235 mm for males and 275 mm for females
Tail length: 118 mm for males and 135 mm for females

Painting dimensions: 340 x 215 mm

Reconstruction: no specimens were available for study so the form was based on the nearest living relative, *Perameles bouganville.* The colour was taken from written descriptions.

The hairy soles of the hind feet may have been an adaptation for life in an arid, sandy habitat.

in the Northern Territory and from Well 35 on the Canning Stock Route. The most recent specimen was taken at Gahnda, Western Australia, in 1931. There are very few museum specimens and no photos of live animals in existence. Evidence from Aborigines suggests that it may have persisted in the Tanami Desert area until the early 1960s, but it is now presumed extinct.

Precise reasons for its apparent extinction are not clear. However, there are several factors which may have contributed to its decline in some areas. Vegetation throughout its range has been altered by grazing — by stock and rabbits in the eastern half, and by rabbits alone in the western half. Introduced predators such as foxes and feral cats may also have been a factor. The cessation of Aboriginal fire regimes has resulted in decreased diversity and increased uniformity of vegetation, and has made possible the devastation caused by larger uncontrolled fires. Medium-sized animals such as the desert bandicoot appear to be the ones most adversely affected by these kinds of changes.

A number of local Aboriginal names have been recorded for the desert bandicoot. These include *mulgaruquirra* (Alice Springs), *iwurra* (Charlotte Waters) and *walilya* (Warburton region).

DESERT RAT-KANGAROO

(Caloprymnus campestris)

FROM
The Mammals of Australia,
John Gould, 1848-1863
Gould's name for the mammal:
Plain-loving Jerboa Kangaroo,
Bettongia campestris.

A distinguishing feature was its very large, thick upper lips, which swelled outwards beyond the level of the nose.

Last sighting 1935

Vulnerable	*Could become endangered because few in numbers or limited in distribution.*
Endangered	*Likely to become extinct unless action is taken.*
Critical	*Only a few hundred individuals survive, or only a single small population known.*
Extinct	*Not sighted for more than 50 years.*

Since European settlement in Australia, the desert rat-kangaroo has had a most elusive and enigmatic history. Although it was described in 1843 by John Gould, from only three specimens sent by the governor of South Australia, Sir George Grey, no more evidence for its continued existence was to emerge for nearly a century. It was rediscovered and recognised by Europeans in 1931 (Aborigines were, almost certainly, well aware of its existence) but disappeared again shortly after.

The desert rat-kangaroo, or *oolacunta* as it was known to local Aborigines, was a small rat-kangaroo species that weighed between 637 and 1060 grams, depending upon sex and age. Females were evidently heavier than males. The head and body measured 254-277 mm, with the tail adding a further 297-377 mm. The animal's soft, dense fur was an overall sandy buff colour which was darker in the centre of the back and the top of the head and faded to a lighter shade at the sides. The chin, sides of the chest and stomach were creamy white. Both the front and back legs were redder, and the hands and feet paler, than the rest of the body. The tail was covered with pale, yellowish hairs. Both sexes had a patch of naked, thickened skin, about 50 mm long and 20 mm wide, between the forelimbs, but its function remains unknown. Another distinguishing feature of the desert rat-kangaroo was its very large, thick upper lips, which swelled outwards beyond the level of the nose.

As the desert rat-kangaroo re-emerged for such a short time, little work was done on its biology and behaviour. It was, moreover, not possible to make observations of undisturbed animals. Virtually everything we know about it comes from the studies carried out by Hedley Finlayson between 1931 and 1935. The female had a deep pouch containing four teats but appears to have had only a single infant at a time. A few adult females collected in December, and one in August, had large, furred and partly independent young while one had a much smaller, naked pouch young. Three females taken in June had pouch young in different stages of growth. It seems from this evidence that the breeding season was quite long, or perhaps it bred only after good rains. The animal appears to have been solitary except for females accompanied by unweaned young.

Unlike many other small mammals of the arid zone it did not burrow but dug a shallow hole about 10 cm deep and 25 cm in diameter. In this it built a flimsy nest of leaves and grasses which in exposed situations was thatched over with twigs and grass stems. More

DESERT RAT-KANGAROO

Caloprymnus campestris

Head and body length: 254-277 mm
Tail length: 297-377 mm

Painting dimensions: 430 x 280 mm

Reconstruction: as only a flat skin was available, and it could only be used for colour reference, the form was built up from research into the nearest existing relative, the rufous rat kangaroo, plus very old photographs of 'Oolacunta' by Finlayson.

When travelling at speed, the right foot was 7 cm in front of the left, with the left striking the ground obliquely.

commonly the nest was built under a bush. Aborigines would note the direction in which the animal's observation hole opened and creep up on the opposite side and grab the unsuspecting creature. When Finlayson went in search of the desert rat-kangaroo in 1931 he did not believe his Aboriginal informant until he observed him catching the animal in this way. Finlayson had to travel 19 km, on three different horses, before he could tire and catch his own specimen. He noted its amazing endurance and speed in such arid conditions. Aborigines could also locate this animal by following its distinctive tracks. Its tracks showed that when travelling at speed, the right foot was 7 cm in front of the left, with the left striking the ground obliquely.

The animal was largely nocturnal and was thought to eat the green parts of plants rather than roots. However, its well-developed foreclaws suggest that it may have dug and eaten more underground food than Finlayson thought. A recent anaylsis of the contents of the digestive tract of a specimen kept pickled in the Museum of Victoria for over 50 years found that the remains of beetles were present throughout. This suggests a reliance on insects that was previously unsuspected, although the rufous bettong, its near relative in eastern Australia, eats considerable numbers of insects. Finlayson also noted that while it seemed independent of water it did not favour the sandhills where more succulent plants grew. However, unusual conditions at the time may have considerably affected the animal's behaviour; for good rains that fell some months before Finlayson's expedition had broken a drought of several year's duration.

There is no locality recorded for the three specimens described by Gould in 1843. Gould noted that they were taken from 'the stony and sandy plains of the interior of South Australia, partially clothed with scrub'. According to Finlayson, these earlier specimens must have come from an area somewhere south of the site of the animal's eventual rediscovery, as land exploration by Europeans in the 1840s had not reached this far north. It is not really surprising that it took so long for the species to re-emerge. Field work in this part of Australia was fraught with difficulties; Finlayson travelled more than 1300 kilometres by car and 500 by horse, crossing Sturt's Stony Desert in extreme temperatures in order to reach his eventual collection point.

The animal was first brought to the attention of Professor F. W. Jones when Mr L. Reese from Appamunna Station sent a description in 1924. Presumably this data arrived too late to be mentioned in Jones' book of the same year-where he suggests that

it was probably extinct. It was not till Reese sent a specimen to Finlayson in 1931 that an expedition was mounted. This resulted in several new specimens, as well as photographs and notes. Most of these specimens were taken on flats which represent a transitional stage between the true gibber plains and loamy flats. The sparse vegetation of these areas includes saltbush, emu bush and the occasional clump of stunted corkwood trees. The area is hot and dry and subject to drought, and winter nights are frequently frosty. According to Finlayson the animal rarely frequented the sandhills. Finlayson found it hard to believe his Aboriginal informants when they told him that this was the best place to obtain the desert rat-kangaroo: '. . . anything less promising than this stony, shelterless plain in the blaze of midsummer, would be difficult to imagine. But the blacks were right', he wrote in 1935.

Desolate areas such as this one, around Lake Eyre, were once home to the desert rat-kangaroo. It appears to have been a solitary creature apart from females accompanied by unweaned young. Unlike many other small mammals of the arid zone, it did not burrow but dug a shallow hole about 10 cm deep and 25 cm in diameter. In this it built its nest of leaves and grass.

(P. W. Menkhorst/NPIAW)

The desert rat-kangaroo was eventually found to inhabit a small area of northeastern South Australia and the adjacent part of Queensland, east and northeast of Lake Eyre, in the drainage of the lower Dimantina River and Cooper Creek. Fossils attributed to this species have been found at Lake Menindee, New South Wales. Fossils found in far southeastern Western Australia, which some scientists have attributed to this species, are now thought to belong to yet another kind of extinct desert rat-kangaroo.

The desert rat-kangaroo has not been recorded since 1935. Although Gould made no mention of its rarity, it seems that the animal may never have been common. It has been suggested that its numbers were abnormally high in the 1930s because abundant vegetation was available after unusually good rains. Unfortunately there is little likelihood that the animal will make another amazing reappearance. In good seasons most of the area is now subject to cattle and rabbit grazing which leads to soil and vegetation degradation. The presence of foxes and feral cats has further lessened the chances that this species still exists.

EASTERN HARE-WALLABY

(Lagorchestes leporides)

FROM
The Mammals of Australia,
John Gould, 1848-1863
Gould's name for the mammal:
Hare Kangaroo,
Lagorchestes leporoides.

The eastern hare-wallaby's likeness to a hare is accentuated by its short head and the length of its ears.

Last sighting 1890

Vulnerable	*Could become endangered because few in numbers or limited in distribution.*
Endangered	*Likely to become extinct unless action is taken.*
Critical	*Only a few hundred individuals survive, or only a single small population known.*
Extinct	*Not sighted for more than 50 years.*

The eastern hare-wallaby was scientifically described by John Gould in 1841, and was the first of the four hare-wallaby species to become known to Europeans. The generic name that he gave the whole group, *Lagorchestes* (which is Greek for 'dancing hare'), is particularly appropriate to the eastern hare-wallaby. Hare-wallabies are characterised by a hairy muzzle and resemble the European hare in size and in the texture and colour of their fur. The eastern hare-wallaby's likeness to a hare is accentuated by its short head and the length of its ears. Gould also thought that some of their habits were similar to those of hares.

One stuffed specimen has been recorded as having a head and body length of 450 mm, with a tail 320 mm long. There are no recorded weights for this species as it became extinct before weighing became standard scientific practice. The eastern hare-wallaby had an overall grizzled brown colour on its top side. The guard hairs were dark grey with white tips, while the individual hairs of the long, soft underfur were dark grey at the base with broad brown tips. A faint, pale hip-stripe could be seen in some individuals and a more prominent black patch occurred on the elbow. The hair on the feet was very short and was of a greyish white colour, and the central hind claw was very long. The tail, too, was sparsely haired, being brownish grey on top with white on the sides and underneath. There was also an indistinct reddish patch around the eyes and the nose was covered with short brown hairs with a black, naked patch on the tip. Gould noted a considerable degree of variation in colour between individuals.

Little is known about the eastern hare-wallaby's reproduction and behaviour. What has been recorded is confined to the writings of the nineteenth-century naturalists John Gould and Gerard Krefft. Krefft noted that the females had four teats in the pouch and produced one young at a time. Almost all other kangaroos share these features. Gould mentioned the animal's fleetness over short distances, and described as extraordinary its powers of leaping. While in the field Gould observed an eastern hare-wallaby running away from dogs: instead of branching either right or left to avoid colliding with him, it 'bounded clear over my head' (Gould 1863). Krefft claimed it could leap more than eight feet (2.4 metres) high. Gould thought it was solitary in its habits, noting that he found it: '. . . sitting close in a well-formed seat under the shelter of a tuft of grass on the open plains'. Krefft mentioned the solitary nature of the animal, noting that it was strictly nocturnal and slept under saltbush or in

EASTERN HARE-WALLABY

Lagorchestes leporides

Head and body length: 450 mm
Tail length: 320 mm

Painting dimensions: 400 x 220 mm

Reconstruction: a mounted specimen in reasonable condition was viewed and sketched, and coat and ear details were filled in by following written descriptions.

The eastern hare-wallaby was once abundant on the inland plains of New South Wales, where vast numbers of sheep now graze. Its scientific name means 'dancing hare'. It was the first type of hare-wallaby to become known to Europeans.
(NATFOTO/ANT Photo Library)

Gould mentioned the animal's fleetness over short distances, and described as extraordinary its powers of leaping.

any other sheltered locality during the day. Krefft found it easy to tame and fed tamed individuals on biscuit, bread or boiled rice. In his own distinctive style Krefft often commented upon the culinary attributes of various mammals, leaving us the only descriptions of how some now-extinct species tasted. In the case of the eastern hare-wallaby he noted, in 1866, that: 'the flesh is delicious, in fact some of the best meat I ever tasted'.

The eastern hare-wallaby declined rapidly in the second half of the nineteenth century. In the 1840s Gould reported it to be tolerably abundant in all the plains of South Australia, particularly those situated between the Murray River and the mountain ranges. He also recorded it in New South Wales from the Liverpool Plains and the vicinity of the Namoi and Gwydir Rivers. According to Krefft, in 1856-1857 it was common upon the level country between the Murray and Darling Rivers, where the local Aborigines knew it as the *turatt*. It inhabited inland woodland areas and plains with long grass and shrubby ground vegetation. Today the eastern hare-wallaby is very poorly represented in museum collections, and has not been recorded or collected since a specimen was taken near Booligal, New South Wales, in 1890. A combination of factors — including overstocking with sheep up to the 1890s, invasion by rabbits during the 1880s and severe drought — may have contributed to its demise. However, the precise reason for its apparent, and tragic, extinction remain a mystery.

GOULD'S MOUSE

(Pseudomys gouldii)

FROM

The Mammals of Australia,

John Gould, 1848-1863.

Gould's name for the mammal:

White-footed Mouse,

Mus gouldi.

Its small size and rodent appearance could easily lead to its being mistaken for an introduced rat pest and destroyed.

Because Gould's mouse had a recorded history of less than 20 years, this typically mouse-shaped native rodent is one of the least known of all Australian mammals. Described scientifically by Waterhouse in 1839, it was last collected alive by Gerard Krefft during the Blandowski Expedition of 1856-57.

The head and body of this elusive creature measured 100-129 mm, and the tail was 90-100 mm long, making Gould's mouse intermediate in size between a rat and a mouse. Its weight has been estimated at about 50 grams. Waterhouse's original description is partly used here. The general colour of its soft fur was yellow-brown above with numerous long black hairs interspersed. The hairs on the back were a deep lead colour at the base, pale ochre towards the tip and dusky-coloured on the extreme tip. This gave the animal a dark, variegated appearance. The underside was a greyish, buff-white. The large and slightly pointed ears were brown which, except at the tip, was interspersed with yellow hairs. The thin tail was brownish on top and yellowish white underneath. Whiskers were long and brown. From Gould's and Waterhouse's descriptions it appears that there was a degree of colour variation within the species. In the past confusion between it and other members of its genus resulted in some erroneous identifications. It can, however, be distinguished from other, closely related species by its long and slender white feet as well as by some features of the skull and skeleton.

As a result of its early demise, extremely little is known about the biology and behaviour of Gould's mouse. It appears to have been a social species; Gilbert recorded that it lived within burrows in 'small families of from four to eight in number, inhabiting the same burrow, and even the same nest of dried soft grasses' (Gould 1863). Extensive burrows were constructed about 15 cm below the surface. Gilbert, who collected in Western Australia, noted it was found in loose soil on the sides of grassy hills, while Gould recorded its preference for the sandhills and plains of the interior. It seems that it was frequently preyed upon by owls. Evidence for this are remains found in pellets, regurgitated by owls, containing indigestible parts of their meals.

The first specimen to be found and described was recorded as coming from New South Wales, probably the upper Hunter River district or the Liverpool Plains, as well as some from the Moore River in Western Australia. Other specimens were collected during the Blandowski Expedition to the junction of the Murray and Darling

Last sighting 1856-57

Vulnerable	*Could become endangered because few in numbers or limited in distribution.*
Endangered	*Likely to become extinct unless action is taken.*
Critical	*Only a few hundred individuals survive, or only a single small population known.*
Extinct	*Not sighted for more than 50 years.*

GOULD'S MOUSE

Pseudomys gouldii

Head and body length: 100-129 mm
Tail length: 90-100 mm

Painting dimensions: 322 x 164 mm

Reconstruction: the colour was taken from written description, but the form was based on a single poor specimen, and research on close relatives such as *Pseudomys australis*.

The sandhills and plains of inland Australia are covered in wildflowers after good rains. Gould's mouse probably inhabited countryside such as this near Tibooburra, New South Wales.
(Otto Rogge/ANT Photo Library)

Gould's mouse appears to have been a social species that lived within burrows in small families.

Rivers in 1856-57. These proved to be the last living specimens collected. Sub-fossil remains have also been recorded from Lake Victoria in southwestern New South Wales and Chamber's Gorge in the Flinders Ranges, South Australia. The material collected from the Flinders Ranges is from owl pellets accumulated in caves. The remains of Gould's mouse are common in these deposits and this suggests that it may once have been common in this area. Although the precise age of these owl pellets is not known, similar-looking pellets contain remains of the introduced house mouse. This indicates the fairly recent presence of Gould's mouse in the area.

There has been extensive habitat disturbance as a result of stock grazing and land clearance in the areas where Gould's mouse was collected last century. Although the animal has been presumed extinct, its small size and rodent appearance could easily lead to its being mistaken for an introduced rat pest and destroyed. There is only a very remote chance that some time in the future this animal, known with certainty for less than two decades, will reappear. Authors who have pronounced it extinct will surely not be sorry to be proved wrong.

LONG-TAILED HOPPING-MOUSE

(Notomys longicaudatus)

FROM
The Mammals of Australia,
John Gould, 1848-1863.
Gould's name for the mammal:
Long-tailed Hapalotis,
Hapalotis longicaudata.

It was very partial to the mounds thrown up by the burrowing bettong and the bilby.

This very pretty rodent has been found alive by Europeans on only four occasions, and then from widely separated localities. It was first described scientifically by John Gould in 1844. It was larger than most other hopping-mice; its head and body were 125 mm long and its tail, which was considerably longer than the body, measured an impressive 180 mm. Its weight has been estimated at 100 grams. The dense, very soft fur was tawny brown on the back, grading to grey-white under the limbs. Its large ears were dark brown and naked. Characteristic of hopping animals, it had long hind feet. It was distinguished from other hopping-mice by its size and its long bi-coloured tail, which was brown above and white below. At the end of the tail was a black brush which, according to Gould, may have been white at the extreme tip. The males also possessed an oval neck gland with a raised margin, but little is known about its function.

Like many of Australia's other lost mammal species, this animal disappeared before scientists could collect sufficient information about its biology and habits to enable them to determine its ecological niche. Some of the limited information we have about the species was supplied by the collector John Gilbert and reported by Gould. The long-tailed hopping-mouse must have been reasonably common in the area of Western Australia where Gilbert collected his specimens, because he commented that it '. . . is less destructive to the sacks and bags of the store-room, but, like the *H.mitchelli* [Mitchell's hopping-mouse], is extremely fond of raisins . . .' (Gould 1863). Gilbert also noted that the long-tailed hopping-mouse resembled the smaller Mitchell's hopping-mouse in general appearance but '. . . differs considerably in its habits from the Djyrdow-in [local Aboriginal term for Mitchell's hopping-mouse] for while [the latter] burrows in sandy districts, the favourite haunt of the present species is a stiff clayey soil. It is also very partial to the mounds thrown up by the Boordee [burrowing betong] and the Dalgoitch [bilby] . . .' (Gould 1863).

Fossils, as well as the few records of living animals, suggest that the long-tailed hopping-mouse was once numerous in arid and semi-arid country throughout Australia. Vegetation included eucalypt and acacia woodlands, hummock grassland and low shrubland. Gould's original specimens were collected by Gilbert from the Moore River near Perth in Western Australia. A single specimen was collected by the explorer Captain Charles Sturt in 1845 in the Coonbaralba Range about 140 kilometres from Laidley's Ponds near Broken Hill, New

Last sighting 1901

Extinct	*Not sighted for more than 50 years.*
Vulnerable	*Could become endangered because few in numbers or limited in distribution.*
Endangered	*Likely to become extinct unless action is taken.*
Critical	*Only a few hundred individuals survive, or only a single small population known.*
Extinct	*Not sighted for more than 50 years.*

LONG-TAILED HOPPING MOUSE

Notomys longicaudatus

Head and body length: 125 mm
Tail length: 180 mm

Painting dimensions: 318 x 218 mm

Reconstruction: a well-preserved specimen was used as a basis for the sketches, with details such as the ears added from research into other existing species of hopping mice.

The long-tailed hopping-mouse once inhabited plains such as this across southern Australia. It is now extinct but is used to be numerous in semi-arid country, among the eucalyptus and the acacia woodlands, as well as hummock grassland and low shrub lands.

(K. Johnson/NPIAW)

South Wales. The specimen was erroneously thought to be a separate sub-species and was named in honour of Sturt as *Notomys longicaudata sturti*. Sturt described it as a rat-like diminutive kangaroo and noted that it was known as *talamba* by the local Aborigines. Several specimens were collected on the Burt Plains, north of Alice Springs, Northern Territory, during the Horn expedition of 1894. The last authenticated specimens were collected from Barrow Creek, Northern Territory, in 1901 during the Spencer and Gillen expedition. A skull fragment found in 1977 in owl pellets from a cave 96 km south-southwest of the Granites, and an earlier fragment from the Tanami Desert area in the Nothern Territory, may be from the long-tailed hopping-mouse. Similar remains in owl pellets have also been found at Chamber's Gorge in the Flinders Ranges, South Australia. However, even if these did prove to be remains of the long-tailed hopping-

mouse, the owl pellets in which they were found may well be decades old. Unfortunately, the species is now considered extinct.

The long-tailed hopping-mouse is one of only three large hopping-mice previously found in Australia's arid areas. All have disappeared rapidly. The little information we have about the long-tailed hopping-mouse is a great deal more than what we have for the others, the short-tailed hoppping-mouse, which is known from only two specimens and an unnamed species only known from fossils. A devastating chain of events must have been triggered when European settlement started changing the delicate desert ecology.

Sturt described it as a rat-like diminutive kangaroo.

PIG-FOOTED BANDICOOT

(Chaeropus ecaudatus)

FROM
The Mammals of Australia,
John Gould, 1848-1863.
Gould's name for the mammal:
Chestnut-eared Chaeropus,
Chaeropus castanotis.

The first known specimen had lost its tail, and this was mistakenly thought to be a feature of the species.

The pig-footed bandicoot was undoubtedly the strangest of all bandicoots. First collected in 1836 by surveyor-general Thomas Mitchell, it was scientifically described by William Ogilby in 1838. It was an extremely delicate and graceful animal, about 230-250 mm long, with a tail that added another 100-150 mm. It had a coarse coat which was orange-brown above and fawn below — both mixed with slaty grey. Its long slender tail had a black crest at the tip. The second half of the scientific name, *ecaudatus.* means tail-less. The first known specimen had lost its tail, possibly in fights with neighbours, and this was mistakenly though to be a feature of the species. When the distinguished zoologist Gerard Krefft went in search of the animal in 1857, he showed Aborigines a drawing of the tail-less pig-footed bandicoot. He was rewarded with a large number of more common bandicoots, 'minus the tail, which to please me had been screwed clean out' (Krefft 1866).

The pig-footed bandicoot had fairly long ears and a sharp, pointed nose that was hairless at the extreme tip. Its limbs were long and slender and its forefoot had only two functional toes, which gave it a superficial resemblence to a pig's foot — hence the animal's common name. The fourth toe of the hind foot was much larger than the others, and was the only functional toe on each hind foot.

Because of its early disappearance from the areas settled by Europeans, not much has been recorded about the biology and habits of this species. Mitchell made only rudimentary observations, and Krefft was the first white person to properly observe and investigate the pig-footed bandicoot in its natural habitat. Krefft observed the animal running away, commenting that 'it progressed like a broken down hack in a canter', apparently dragging its hindquarters behind it. Animals which Krefft kept in a cage for sketching and observation constructed a completely covered nest with barley grass and dried leaves. They ate lettuce, barley grass, bread and bulbous roots for six weeks before they were killed to make scientific specimens. Krefft reported that dung, which he observed while hunting and which he attributed to this species, was very dry and entirely composed of grass. Local Aborigines verified that this constituted the diet of the pig-footed bandicoot. Recent examination of the contents of the digestive tract of two specimens collected last century near Alice Springs, and held in the Museum of Victoria, have confirmed Krefft's observations. The faecal pellets found in the intestine of the preserved

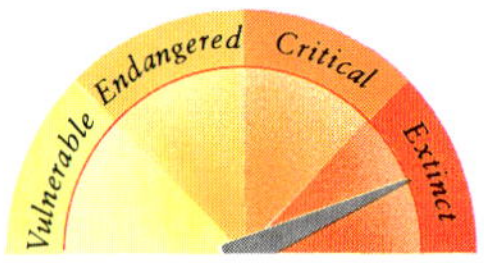

Last sighting 1907

Vulnerable	*Could become endangered because few in numbers or limited in distribution.*
Endangered	*Likely to become extinct unless action is taken.*
Critical	*Only a few hundred individuals survive, or only a single small population known.*
Extinct	*Not sighted for more than 50 years.*

PIG-FOOTED BANDICOOT

Chaeropus ecaudatus

Head and body length: 230-250 mm
Tail length: 100-150 mm

Painting dimensions: 365 x 275 mm

Reconstruction: after viewing the only stuffed specimen available to her, which is in a very poor state, the artist relied on knowledge of the form of relatives like the bilby, and study of earlier illustrations of *Chaeropus ecaudatus*.

It squatted in the open with its ears laid back like those of a rabbit.

specimens were ovoid in shape, 1.5 cm long and 1.0 cm wide, and consisted almost entirely of finely chewed-up grass. Krefft confessed to eating some of his specimens; his appetite having 'more than once over-ruled [his] love for science'. He claimed that eating the plant pigface for 24 hours dampens the ardour of any naturalist, and that pig-footed bandicoots make a tasty repast.

The female pig-footed bandicoot had a backwards-opening pouch with eight teats, although no more than two young at one time have been reported. Early reports of females captured with young suggest a breeding season and birth around May and June.

Although, according to several reports, the pig-footed bandicoot was active in daylight, it was largely nocturnal. Explorers Mitchell and Charles Sturt both reported that, when disturbed, it took refuge in a hollow tree. There were also reports that the animal dodged among the saltbush when hollow trees were not available, and that it squatted in the open with its ears laid back like those of a rabbit. From the scant records that are available it seems that the pig-footed bandicoot was a plains dweller with a preference for open woodland where there was an understory of grass and shrubs.

It is certain that the pig-footed bandicoot had an extensive range. It was found in the southern parts of the Northern Territory as far north as Barrow Creek. In New South Wales it was known from the far west to the Murray-Murrumbidgee junction and was recorded, as well, in the northwest corner of Victoria. In South Australia it occurred from the Musgrave and Gawler Ranges to the Cooper Creek and Lake Eyre basins, and Ooldea. In Western Australia it was known in the southern districts. In the middle of last century Sturt also recorded its presence near the common border of Queensland, New South Wales and South Australia.

In this century no specimens of the pig-footed bandicoot have been taken in the Northern Territory, Victoria, New South Wales, Western Australia or Queensland. The last record for New South Wales was in 1857 and for Western Australia in 1843. The last specimen that was recovered was from South Australia in 1907. There was a report of one being killed near Lake Eyre in 1920 and the last recorded sighting of the animal from South Australia was by Pitjantjara men in the Musgrave Ranges in 1926. However, evidence from Aborigines in the Northern Territory suggests that it may have survived until the 1960s in the Tanami Desert and in areas to the west of this region. Few specimens of the animal remain; the original tail-

less specimen was lost from the Australian Museum some time in the last century.

It appears that, even before European settlement, the pig-footed bandicoot was a rare animal within its range. At least one group of Aborigines, the Kukata, believed that the pig-footed bandicoot, which they knew as *wilalya,* had always been rare.

Fossils show that the pig-footed bandicoot's range once extended further to the east and west than is suggested by sightings and collections made in the nineteenth century. Perhaps the range had in fact decreased; perhaps, before it became totally extinct, the species was missed in many areas because it was so rare and because it ranged so widely and was collected by so few people. The truth may never be known. Pig-footed bandicoots were certainly scarce even before European stock, foxes and cats invaded their habitat. The species is now presumed extinct.

The pig-footed bandicoot favoured open woodland with an understory of grass and shrubs such as this at Brookfield Conservation Park, South Australia. The bandicoot was rare even before European settlement.
(T P Morley/NPIAW)

THYLACINE

(Thylacinus cynocephalus)

FROM
The Mammals of Australia,
John Gould, 1848-1863.
Gould's name for the mammal:
Thylacinus,
Thylacinus cynocephalus.

The thylacine was a marsupial and had some curiously kangaroo-like features.

Few animals in Australia's European history have captured the public imagination as has the thylacine, otherwise known as the Tasmanian tiger. Perhaps its carnivorous habits, uncanny similarity to a large dog, and the fact that its extinction was so tantalisingly recent — and indeed is still doubted by many people — have added to its attraction.

Despite its high public profile not a great deal is known about the biology of the thylacine, for there was no zoologist on the staff at the Tasmanian Museum during the period of its rapid decline, and the University of Tasmania did not establish a chair in zoology until 1910. What has been gleaned, however, is fascinating. In spite of its superficial resemblence to a dog, the thylacine was a marsupial and had some curiously kangaroo-like features. These included a tail that was thick at the base and very stiff, a large pouch, and a strip of naked skin between the pad and heel of the hind foot. This latter feature may sometimes have appeared in tracks as the thylacine is reported to have stood on occasion with the heel on the ground in an erect, kangaroo-like stance. The tail was originally thought to be so stiff that if it was held the animal was incapable of turning to bite. The most distinctive feature of the thylacine were dark brown stripes — there were between 13 and 19 of them — that extended down its tail on its sandy brown rump. These, and the thick base of the tail, gave the impression that the tail was a continuation of the body.

The thylacine was not a fast runner and probably caught its prey by exhausting it during a long pursuit. It rarely ran, but trotted instead with its rather stiff gait. If forced it would break into a shambling canter. It hunted singly or in pairs, and during its long-distance pursuits probably relied more on scent than any other sense. It had a particularly large gape — around 120 degrees — and prominent canine as well as shearing molar teeth. It is reputed to have preferred the vascular tissue of its prey, such as the nose, throat, liver and kidney, and was said to have readily included domestic stock in its diet. Its prey included Bennett's wallaby, the long-nosed potoroo, the Tasmanian pademelon, the eastern grey kangaroo, and a variety of bandicoots, rats and birds. It was known to eat carrion when live prey was not available, although some authorities dispute this. It was largely silent, its vocalisations being limited to an occasional terrier-like bark when hunting and a series of husky barks when excited in captivity. Although it was mainly nocturnal, it has been sighted

Last sighting 1936

Vulnerable	*Could become endangered because few in numbers or limited in distribution.*
Endangered	*Likely to become extinct unless action is taken.*
Critical	*Only a few hundred individuals survive, or only a single small population known.*
Extinct	*Not sighted for more than 50 years.*

THYLACINE

Thylacinus cynocephalus

Head and body length: 851-1181 mm
Tail length: 381-610 mm

Painting dimensions: 464 x 239 mm

Reconstruction: the markings were based on one of several mounted specimens in the Australian Museum, New South Wales; the form was based on descriptions and old black and white photographs taken from film footage.

The thylacine was not a fast runner and probably caught its prey by exhausting it during a long pursuit.

moving during the day and captive animals have been recorded basking in the sun.

Adult thylacines could weigh anything between 15-35 kg. Indeed, recorded weights of live animals are so few that large males may have weighed considerably more. There was an extended breeding season from winter to spring. The females had four teats in a backwards-opening pouch that enlarged and hung down low as the young grew. The maximum number of young was four, although the average litter was almost certainly smaller, and when the 'pups' became too large to be carried they were left in a cave or shelter. After weaning they accompanied the mother until they were well grown and able to hunt independently. Another curious feature of the thylacine was the male's pouch. This was really a cup-like depression, bounded by a flap of skin in front and at the sides, which enclosed the bare and recessed scrotum. The tip of the penis, like that of some other marsupials, was forked.

Not much is known about the lifespan of the thylacine although a captive individual lived in the London Zoo for nearly eight and a half years. As it was probably at least a year old when it was obtained, it would have been more than nine years old when it died. A second specimen lived for 12 years at Beaumaris Zoo, Tasmania. It had arrived as a pup.

Despite its wide distribution in Tasmania in pre-European times it is unlikely that it was ever an abundant species. This is partly because of its position in the food chain, for warm-blooded carnivores are always many times rarer than the species they prey upon. The decline to extinction of the thylacine was precipitous, and the reasons for it are still disputed. Some researchers have postulated that a disease was instrumental in reducing its numbers, and that the decline was accelerated when the thylacine came into competition with domestic dogs, and was hunted by humans. There is, however, little evidence for this disease hypothesis.

The thylacine was found throughout Tasmania, except the rainforests of the southwest. Open eucalypt forest such as this near Hobart was prime habitat. *(Jutta Hösel/ANT Photo Library)*

Approximatley 4000 years ago the thylacine was very widespread throughout the island of New Guinea and most of mainland Australia, as well as Tasmania. Its extinction coincides closely with the arrival of the dingo in Australia and the wild dog in New Guinea. Because dingoes never reached Tasmania most scientists see this as the primary reason for the thylacine's survival there. The most recent, well-dated occurrence of a thylacine on the mainland is a radio-carbon dated fossil from Murray Cave in Western Australia, which is around 3100

years old. A mummified carcass has also been found in Western Australia, and it is radio-carbon dated to be less than 5000 years old. A thylacine upper arm bone from the Kimberley region of Western Australia has been estimated to be 0-80 years old. However, this date is very suspect. Further interesting evidence for the previous presence of thylacines on the mainland includes Aboriginal rock paintings of a striped animal — almost certainly a thylacine — in the Kimberley region of Western Australia and in the Northern Territory.

By the time Europeans encountered the thylacine it was restricted to the island of Tasmania. Indeed the first reference to a thylacine may have been made by the Dutch explorer Abel Tasman, who reported seeing some strange footprints when, in 1642, he landed near present-day Dunalley. The first definite reference is that of Paterson who reported in March 1805 that his dogs were being killed by an animal 'of carnivorous and voracious tribe' near Yorktown on the Tamar River. There is also mention shortly after this in the Reverend R. Knopwood's diary for June 1805. Knopwood was the official padre to the colony and wrote of a 'large tyger'. The first scientific description was by Harris in 1808.

Despite its wide distribution in Tasmania in pre-European times it is unlikely that it was ever an abundant species.

Even today museum curators and other scientists throughout Australia regularly receive reports of sightings of thylacine-like animals, a favourite locality for such reports being the densely populated Blue Mountains area west of Sydney! Despite these, and frequent media speculation, a number of facts strongly suggest that this unique animal is completely extinct. The last thylacine taken from the wild was shot in the Florentine Valley in 1933, while the very last recorded individual died in Hobart's Beaumaris Zoo in 1936. Because much of the rugged west of Tasmania is impenetrable, many people remain optimistic that thylacines still survive there. However, the dense rainforest area of southwestern Tasmania is the one habitat that appears not to have supported thylacines. None of the numerous expeditions and searches, which began in 1938, and culminated with the $250 000 search of 1984, has produced a single hair to suggest that thylacines still exist. These included searches by leading Australian naturalists such as David Fleay, who searched for months in 1945-46, and notable thylacine experts such as Eric Guiler, who has searched for years.

It seems that predation by Europeans, firstly by the Van Diemen Land Company and private landowners from 1820, and then as a result of bounties paid by the Tasmanian government from 1888 to 1912, was the primary factor in exterminating the thylacine. The

Tasmanian government bounty was one pound for each adult scalp and ten shillings for sub-adults. Before they were withdrawn in 1912, bounties were collected on 2184 of these rare animals. However, only two individuals were presented for bounty payment in 1909 and none in the last three years of the scheme. The government bounty numbers quoted do not include those thylacines killed for private landowners, who also offered rewards. There seems to have been little public pressure to preserve the species, nor was much concern expressed by the public or scientists at the decline of this spectacular beast. A notable exception was T. T. Flynn (Errol Flynn's father), professor of biology at the University of Tasmania. In 1914 he was sufficiently concerned about the scarcity of the thylacine to suggest that some should be captured and placed on an island.

Tragically, nothing was done. It was not until 1929, with the species on the very edge of extinction, that the Animals and Birds Protection Board passed a motion protecting thylacines only for the month of December, which was thought to be their prime breeding season. This was gazetted in 1930. On July 14, 1936 the thylacine was belatedly declared a totally protected species — which, ironically, it remains to this day. There was never a successful attempt to breed thylacines in captivity despite the large numbers that went through some zoos, particularly those in Hobart and London. When they were caught, however, some females had cubs with them, and a few of these survived. Even when thylacines were declining to extinction, the only possible breeding groups were being broken up to be exchanged for 'rarities' such as zebras and lions, or to be sold for cash.

There was never a successful attempt to breed thylacines in captivity, despite the large numbers that went through some zoos.

The famous naturalist John Gould foresaw the thylacine's demise when he published his *Mammals of Australia* between 1848 and 1863: 'When the comparatively small island of Tasmania becomes more densely populated, and its primitive forests are intersected with roads from the eastern to the western coasts, the numbers of this singular animal will speedily diminish, extermination will have its full sway, and it will then, like the Wolf of England and Scotland, be recorded as an animal of the past'.

Today, all that remains of thylacines, apart from a few photographs, paintings and a short black-and-white film, are some specimens preserved in museums. There is a whole adult and a pouch young preserved in alcohol. Perhaps, at some distant future date, it may be possible to isolate DNA from one of these bodies, and perhaps even to clone a thylacine.

TOOLACHE WALLABY

(Macropus greyi)

FROM

The Mammals of Australia,

John Gould, 1848-1863.

Gould's name for the mammal:

Grey's Wallaby,

Halmaturus greyi.

The toolache wallaby was hunted for its beautiful pelt.

The demise of the beautiful toolache wallaby is one of the most tragic occurences in the history of the European occupation of Australia. Once a common animal within its fairly small range, it seemed to disappear almost overnight. Its attractive coat and swift gait certainly contributed to its extinction, for the toolache wallaby was hunted for its fur as well as for sport, its speed giving excitement in the chase. It was described scientifically by Waterhouse in 1845 under the generic name of *Halmaturus*, and received the second part of its scientific name in honour of Governor Sir George Grey, who sent the first specimen to England.

The elegant, slender, toolache wallaby was perhaps the most beautiful of the larger wallabies. The recorded head and body length for males is 810 mm, with the tail adding another 730 mm. Females are recorded as having a head and body length of 840 mm and a tail length of 710 mm. These are measurements of stuffed individuals. There are no recorded weights. The general colour of the toolache wallaby was pale ashy brown slightly tinted with yellow on the back. The underside was a pale buff-yellow. The back of the ears and shoulders had a reddish tinge while the ear tips were blackish. Between the rump and the base of the tail the back was marked with about a dozen transverse bars which were alternately light and dark grey. These bars resulted from the texture as well as the colour of the fur and may have varied according to the season and the individual. There was also an indistinct whitish hip-stripe. The long slender tail was pale grey and became paler towards the tip which was almost white. There was a crest of long white hairs on the last third of the tail. The forearms and feet were pale fawn and contrasted with the sharply demarcated black hands and toes. The muzzle was long and broad, and bordering the nose was two to three centimetres of black fur. There was a prominent black mark from the nose to the eye, bordered below by a white cheek-stripe which reached almost to the ear. There was a white mark above the eye.

Nothing has been recorded about the reproduction of the toolache wallaby. In historic times it inhabited open country and avoided forested and dense heath areas. Its preferred habitat was a dense, tall sedgeland and grass community that had developed on the clayey soils of plains between parallel, low, consolidated dune ridges. These areas were too wet in winter to carry eucalypt cover. On poorer country the toolache wallaby seems to have been a fairly solitary animal,

Last sighting 1939

Vulnerable — *Could become endangered because few in numbers or limited in distribution.*

Endangered — *Likely to become extinct unless action is taken.*

Critical — *Only a few hundred individuals survive, or only a single small population known.*

Extinct — *Not sighted for more than 50 years.*

TOOLACHE WALLABY

Macropus greyi

Head and body length: 810 mm for males and 840 mm for females
Tail length: 730 mm for males and 710 mm for females

Painting dimensions: 430 x 235 mm

Reconstruction: the anatomy was based on black and white photographs from old film footage of *Macropus greyi*, plus observation and notes made on viewing a flat skin in the Australian Museum, New South Wales.

Groups of these mammals seemed to show a strong attachment to their own areas, returning to them even after repeated disturbance.

but on more developed grasslands it was gregarious, grazing and resting in groups. These groups seemed to show a strong attachment to their own areas, returning to them even after repeated disturbance. A strip of grassland, up to 50 metres wide, that occurred between the edge of the heath and the sedgeland, was a preferred feeding area. This thin strip also provided an ideal setting for the toolache wallaby to display its striking and characteristic turns of speed and its unusual gait. Both historical and contemporary authors have commented on the animal's speed and gait. John Gould, writing in 1863, quotes a Mr Strange: 'I never saw anything so swift of foot as this species; it does not appear to hurry itself until the dogs have got pretty close, when it bounds away like an antelope, with first a short jump then a long one, leaving the dogs far behind it . . . I have had twenty runs in a day with four swift dogs and not succeeded in getting one'. Professor F. W. Jones described its gait as 'two short hops and a long one', and older people living in its previous range state that its gait was erratic, and that it was adept at changing direction rapidly. Furthermore, when chased it could jump to clear a considerable obstacle. The stride was also distinctive, with the greater part of the body extending horizontally. It has been recorded as travelling through complex tunnels in the heath without breaking through the top.

Little has been recorded on other aspects of the toolache wallaby's behaviour. The biologist and explorer Hedley Finlayson recorded an incident in which a female toolache wallaby drove off a wedge-tailed eagle that was attempting to attack its young. The wallaby sprung at the eagle '. . . and with both hind feet struck the peculiar lashing blow common to most macropods'. There were reports of European children raiding wedge-tailed eagle's nests in order to collect infant wallaby scalps for the sixpence bounty payment on all marsupials at that time. As well as eagles, foxes were reported as predators.

At the time of European settlement toolache wallabies appear to have lived in the southeastern part of South Australia — from Lake Albert probably as far south as the southern coast, and eastwards to the adjacent far western part of Victoria. However, no modern specimens are known to have been collected from Victoria. The toolache wallaby had a much wider distribution in late Pleistocene times, approximately 14 000-20 000 years ago. Fossils have been recorded from northwestern Tasmania, Hunter Island in Bass Strait, Kangaroo Island in South Australia, and the Mount Hamilton Lava

Cattle now graze and crops grow where the last toolache wallabies were hunted to extinction in the southeast of South Australia. It is sad to think that this animal, which was described as elegant, slender and the most beautiful of the larger wallabies, is now lost to us.

Cave in Victoria. At that time the climate was cooler, annual rainfall was lower and grasslands were more widespread. By 1910 populations had been reduced to a number of scattered colonies in the area enclosed by Robe, Kingston and Beachport on the coast, and Naracoorte and Penola further to the east. By 1924 only one small group was known to survive — on Konetta Station, about midway between Robe and Penola. Writing about the animal in the same year, Jones suggested that 'any effort to preserve this remnant must be made immediately and with vigour if it is to be of any service whatsoever'.

The Flora and Fauna Board of South Australia decided that a greater chance of protection would be afforded if some of the remnant population was removed to a recently established sanctuary on Kangaroo Island. Unfortunately attempts to do this — in May 1923, and again in 1924 — failed, and the four animals that were captured died either immediately or shortly after capture as a result of being driven too much. The extensive publicity given to these expeditions further accelerated the demise of the species because the survivors of the Konetta group were hunted and killed to obtain trophies! The last known survivor was rescued from dogs on Konetta Station in 1927 and lived in captivity in Robe until its death on 30 June 1939. In spite of numerous and persistent anecdotal reports since then, only one appears to be a record of a toolache wallaby — and that

It was as swift of foot as an antelope — bounding away first with a short jump, then with a long one.

When chased, it could jump to clear a considerable obstacle.

was in 1943. Other reports that have been investigated in the field by biologists have remained unsubstantiated or have been shown to involve another species, commonly the similar-looking red-necked wallaby, *Macropus rufogriseus.*

Although the toolache wallaby was hunted for its beautiful pelt and for sport, the biggest factor in its decline was the extensive opening up of its habitat that resulted from the drainage and clearing of native vegetation between the 1860s and the early 1900s. The animal survived only 85 years of European occupation. Any hope that remnant populations might survive was further diminished with a second period of land clearance in the 1950s. Finlayson, writing in 1927, considered the fox to be the sole factor in its extermination, but although the fox may have contributed to the toolache's destruction, it cannot be blamed entirely for its extinction. Unfortunately, and despite many erroneous reports to the contrary, the chance that this attractive, slender wallaby is not extinct is an extremely remote one.

WHITE-FOOTED RABBIT-RAT

(Conilurus albipes)

FROM
The Mammals of Australia,
John Gould, 1848-1863.
Gould's name for the mammal:
White-footed Hapalotis,
Hapalotis albipes.

Its common name of 'rat' belies the attractiveness of this squirrel-like rodent.

Last confirmed sighting before 1900

Vulnerable	*Could become endangered because few in numbers or limited in distribution.*
Endangered	*Likely to become extinct unless action is taken.*
Critical	*Only a few hundred individuals survive, or only a single small population known.*
Extinct	*Not sighted for more than 50 years.*

Described by Lichtenstein in 1829, the white-footed rabbit-rat, also known as the white-footed tree-rat, had a very short history after European settlement. Its common name of 'rat' belies the attractiveness of this squirrel-like rodent which had a head and body length of 230-260 mm and a tail length of 220-240 mm. Its weight has been estimated at approximately 200 grams. Its most distinctive feature was its long, bushy, bi-coloured tail, which was dark brown above and white below for its whole length. Although it was smaller than a rabbit the white-footed rabbit-rat's colour and large, broad ears gave it a superficial resemblance to its namesake. As all specimens were collected during the first half of last century, most modern descriptions of the animal have come from the work of John Gould, who worked in Australia in the 1840s. Gould described the white-footed rabbit-rat as having long, soft fur. The greater part of this fur was grey at the base with ashy brown tips. On the animal's back, many fine black hairs were intermixed with these brown tips. The whiskers were black and there was a narrow black line around the eye. The underside of the animal, the inner surface of the limbs and the long feet were all white. Gould also noted that there was some colour variation between animals from different regions.

As the white-footed rabbit-rat diminished so quickly in historic times, information on its biology and behaviour is very limited. It was already considered fairly rare when Gould was working, about 150 years ago, but it is hard to ascertain whether this was its natural state or whether factors that resulted in its eventual disappearance were already operating. Writing in 1863, Gould stated that the white-footed rabbit-rat was '. . . strictly nocturnal in its habits, for it sleeps during the day in the hollow limbs of prostrate trees, or such hollow branches of the large *Eucalypti* as are near the ground, in which situations it may be found curled up in a warm nest of dried leaves. More than once have I, after detecting the animal in its retreat, sawn off the hollow limb and secured it without injury'. The only other totally authenticated information recorded about this species is from Sir George Grey, who observed the animal during his term as governor of South Australia, and recorded these observations in letters to Gould. Grey wrote to Gould: 'This animal lives among the trees. The specimen I send to you, a female, had three young ones attached to its teats when it was caught . . . While life remained in the mother they remained attached to her teats by their mouths

WHITE-FOOTED RABBIT RAT

Conilurus albipes

Head and body length: 233-260 mm
Tail length: 220-240 mm

Painting dimensions: 355 x 262 mm

Reconstruction: the form was taken from a specimen sketched and photographed at the Australian Museum, New South Wales, together with research into the other existing species of rabbit rat.

. . . On pulling the young from the teats of the dead mother, they seized hold of my glove with the mouth and held on so strongly it was difficult to disengage them' (Gould 1863). This habit of clinging to the mother with the mouth is quite common in many Australian rat species. It is tragic that these few notes represent the sum of our unequivocal knowledge on the behaviour of one of Australia's most attractive species.

It slept during the day in the hollow limbs of prostrate trees.

Even detailed knowledge of the animal's former range remains unclear as few museum specimens were collected as evidence. The original specimen that was described by Lichtenstein in 1829 is not given a definite locality, but an old painting attests to the species' presence in the Sydney region in the 1790s. It was reported to be not uncommon inland — presumably around Parramatta. Specimens collected by Ogilby in New South Wales were incorrectly identified at the time as stick-nest rats. Gould listed the known range of the white-footed rabbit-rat as being the southeastern portions of Australia generally, including all parts of New South Wales, Port Phillip and South Australia. Specimens from the Darling Downs in southern Queensland were also recorded by Gould. Animals collected in northwestern Victoria by Gerard Krefft in 1856-57 are now thought to have been incorrectly identified as white-footed rabbit-rats. There is an unlocalised account of the white-footed rabbit-rat occurring somewhere in Victoria in the diary of an early settler, Mr H. S. Parris, who travelled from Melbourne to the Murray River in 1839. Pleistocene and recent fossils from cave deposits have been collected from several localities in southeastern South Australia, southern Victoria, eastern New South Wales and southern Queensland. These back up Gould's statements about the species' distribution.

Although no specimens have been collected since the nineteenth century, there have been some more recent anecdotal reports that appear to relate to this animal. One convincing report was collected by the authors in 1985 from Mr Norman Maher, of Deniliquin in southern New South Wales. During the Depression and until the outbreak of World War II, Mr Maher frequently went rabbiting with his terrier in a small area of woodland a few kilometres north of Deniliquin. He recalls that he often encountered an animal he knew as a wood-rat, which he described as being about the size of the water-rat, *Hydromys chrysogaster,* but with a long bi-coloured tail which was brown on top and white underneath, and typical rat-like incisors. He was familiar with the yellow-footed antechinus, which

also occurs in the region, and it was not this species. It was an attractive animal with a reddish foxy tinge on some parts of its coat. Mr Maher's terrier often located these animals in hollow logs, which they had stuffed to a depth of about 60 cm with nesting material, mainly dried grass. They appear to have been partly diurnal, as Mr Maher sometimes observed them to be active by day. He recalls that, before seeing them, he often heard their peculiar 'ssk ssk' call at a distance, and on one occasion found a female with several young. Mr Maher did not see the 'wood-rats' again after he returned from serving in World War II. He suggests that the mass removal of fallen timber from the area to provide heating for the RAAF base there may have caused their decline. It is likely that this is an account of the white-footed rabbit-rat. If so, the very limited information we have about the species has been increased.

Most of the habitat of the white-footed rabbit-rat has been greatly altered by agriculture and fire. The wheat fields of New South Wales and Victoria probably cover some of its best former habitats. *(G. & B. Cheers/ANT Photo Library)*

Although the white-footed rabbit-rat is presumed extinct, information volunteered by members of the public has resulted in the re-emergence of other animals that were also previously thought to be extinct. In some cases this recognition came just in time to forestall a 'second extinction'. Anyone who thinks they recognise an endangered or extinct species, or an animal not previously recorded in a particular habitat, should contact the closest appropriate institution. Amongst all the 'dead end' reports, there may be one that leads to the rediscovery of an 'extinct' species! Unfortunately the white-footed rabbit-rat appears to have been an early casualty of environmental changes that were set in motion by Europeans.

ENDANGERED NATIVE SPECIES

BANDED HARE-WALLABY

(Lagostrophus fasciatus)

FROM
The Mammals of Australia,
John Gould, 1848-1863.
Gould's name for the mammal:
Banded Hare Kangaroo,
Lagorchestes fasciatus.

It seems to obtain the moisture it requires from dew and the water that is contained in its vegetarian diet.

This beautiful wallaby may have been the first member of the kangaroo family to be seen by a European in Australia. William Dampier, that infamous seafarer, observed and wrote of an animal fitting the description of the banded hare-wallaby when he entered Shark's Bay (now Shark Bay) on 6 August 1699. After a brief description his journal states that they were very good meat! The banded hare-wallaby is a dark grizzled colour above with transverse dark bands across its lower back and rump. The fur is greyish white on the inside of the limbs. The animal's head and body length is 400-450 mm and the tail adds another 350-400 mm. Weight ranges from 1.3-2.1 kg, although males occasionally weigh up to 2.5 kg and females 3 kg.

The female usually gives birth in the latter half of summer, although those that fail to breed then may still produce young later in the first half of the year. Females may also have an embryo in reserve if something untoward happens to their young. This leads to the occasional birth as late as September. While females are usually reproductively inactive in October and November, severe summer drought can disrupt this pattern with breeding occurring as soon as the drought is broken. The young spend about six months in the pouch and are weaned at about nine months. Although males and females are capable of breeding when they are only a year old, they usually do not breed until their second year. While some females may produce two young in a year they usually manage to rear only one. John Gould mentioned the comments made by the French naturalists Peron and Lesueur on their voyage to Western Australia in 1800-04. These adventurers were impressed with the way the females attempted to save their young when hunted: '. . . the devotedness with which they sought to save their offspring was truly admirable; although wounded they fled with the young in the pouch, and never left them until, overcome with fatigue and loss of blood, they could no longer carry them; they then stopped, and squatting themselves on their hind legs, helped the young to get out of the pouch by means of the fore-feet, and sought to place them in a situation favourable for retreat . . .' (Gould 1863).

The banded hare-wallaby is nocturnal. Its diet consists of grasses, spinfex and a variety of shrubs and other plants. It seems to obtain the moisture it requires from the dew and water that is contained in its vegetarian diet. The animal's island habitat includes dense *Acacia ligulata* scrub under which are formed runways that the animal uses

Vulnerable	*Could become endangered because few in numbers or limited in distribution.*
Endangered	*Likely to become extinct unless action is taken.*
Critical	*Only a few hundred individuals survive, or only a single small population known.*
Extinct	*Not sighted for more than 50 years.*

BANDED HARE-WALLABY

Lagostrophus fasciatus

Head and body length: 400-450 mm

Tail length: 350-400 mm

This tiny wallaby is the only living relative of Australia's extinct giant short-faced kangaroos, which are known only from fossils.

(Jean-Paul Ferrero/AUSCAPE)

The banded hare-wallaby was common over much of the south of Western Australia, now it is restricted to the scrublands on islands in Shark Bay.
(NATFOTO/ANT Photo Library)

for shelter during the day. Other low-spreading shrubs and bushes may also be used. These thickets proved almost impenetrable to humans, and Gould recorded that the Aborigines used to burn the thickets every three years. Adults seem to live within individual, well-defined ranges. Although females and juveniles maintain largely amicable relations, males exhibit a high degree of aggression towards each other. Fighting seems to occur in competition for food. Males have a yearly mortality rate of 30 per cent, while for females the figure is 25 per cent. This accounts for the higher number of females in the population.

Western Australia: Bernier and Dorre islands in Shark Bay.

Once common over much of southern Western Australia, the banded hare-wallaby is now restricted to Bernier and Dorre Islands in Shark Bay, Western Australia. Fossils — remains from Aboriginal middens — and a single specimen collected in South Australia last century (now housed in the Berlin Museum) show that it once extended across the southern Nullabor Plain into the lower Murray River region of South Australia. The last mainland specimen was obtained in 1906. After the granting of a pastoral lease on Dirk Hartog Island in 1899, the species gradually declined and finally disappeared from the island in the 1920s. Attempts have been made to reintroduce it there. The decline of the banded hare-wallaby may be attributable, at least in part, to the introduction of domestic stock. While the animal is common on the island sanctuaries that now form its last refuge, it remains vulnerable to further habitat disturbance and to the introduction to these islands of exotic predators.

BILBY

(Macrotis lagotis)

FROM
The Mammals of Australia,
John Gould, 1848-1863.
Gould's name for the mammal:
Long-eared Peragalea,
Peragalea lagotis.

Bilbies gather food items in their tongues, indicating that they will store them in their cheeks before eating them.

The bilby was first described scientifically by Reid in 1837. Its scientific name has been changed several times since then. Before *Macrotis,* it has been known widely under the generic names *Thalacomys* and *Perameles.* This graceful bandicoot is quite variable in size and, depending on sex, habitat and age can weigh anything between 800 and 2500 grams. It has a head and body length of 290-550 mm, with the tail adding an extra 200-290 mm. The bilby has long, silky, blue-grey fur with white on the underside, although seasonal molts may change the coat length and the colour to a fawn-grey. Its ears are long, largely naked and rabbit-like, and it has an elongated muzzle, the last 20 mm of which are flesh-coloured and naked. In some specimens there is a faint indication of bars across the thigh fur. The tail is first grey near its base, and then black, and ends in a sharply defined white tip. A horny spur protrudes beyond the hairs at the extreme tip and there is a crest on the upper surface of the tail. As the bilby moves with its cantering gait it often carries its tail like a stiff banner. When the animal is cantering the hind legs move together and the front legs alternately. It usually has white feet and may have blackish tips on the claws. Some South Australian specimens have black feet.

Despite its endangered status the bilby is quite well known scientifically. Field studies have been carried out on wild populations in the Northern Territory as well as on captive populations. The bilby seems particularly well adapted to its arid environment. It has been shown to have a low metabolic rate with a low water turnover. Because there is little surface water available in its habitat it has developed the capacity to get much of its water from food.

The bilby has strong forelimbs and claws and it uses these to good advantage when digging for food and when burrowing. Although most bandicoots do not make burrows, bilbies dig burrows with a relatively steep spiral to a depth of up to 1.8 metres and length of three metres. These burrows, which usually have single openings, are sited on claypans, loam flats or gibber plains rather than sand hills. The entrance is often next to a termite mound or some shrubs, and is left open. However, when it is at home, the bilby blocks the entrance with soil which extends for some distance into the burrow. If an intruder disturbs this soil the bilby, in an often futile effort to escape, will dig furiously to extend the burrow.

Bilbies are basically omnivorous. They eat insects and their larvae, seeds, bulbs, fruit, fungi and, in captivity, meat and small vertebrates.

Vulnerable	*Could become endangered because few in numbers or limited in distribution.*
Endangered	*Likely to become extinct unless action is taken.*
Critical	*Only a few hundred individuals survive, or only a single small population known.*
Extinct	*Not sighted for more than 50 years.*

BILBY

Macrotis lagotis

Head and body length: varies between 290-550 mm
Tail length: 200-290 mm

The bilby has long silky, blue-grey fur with white on the underside, but this can change with seasonal molts. The coat becomes shorter, and changes to a fawn-grey.

This species is different from most bandicoots in that it excavates extensive burrows in central Australia.

(Babs & Bert Wells/NPIAW)

The bilby favours areas with very little vegetation. Its burrows have single openings often sited on claypans. The burrows have a relatively steep spiral to a depth of up to 1.8 metres.
(K. Johnson/NPIAW)

Captives have been observed to gather food items in their tongues, indicating that they will accumulate and store them in their cheeks before eating them. Sometimes seeds may be licked individually from the ground so that soil and sand are consumed along with them. Sand can comprise between 20 and 90 per cent of the content of bilbies' dung pellets. Bilbies do not have acute sight — smell and hearing are the main senses used in locating food. When digging for food the bilby leaves many shallow holes up to 10 cm deep from which soil has been scattered on all sides. It disperses material with kicks of the hind limbs.

Bilbies seem to be strictly nocturnal, and do not even venture out at dawn or dusk. Because of this they are rarely seen in their habitat and scientists must often use indirect clues such as burrows, tracks, scratchings and droppings to detect their presence. Most bilbies are caught by digging up their burrows.

The female bilby has eight teats in a backwards-opening pouch, but usually has litters of no more than two. Mating occurs above ground in captivity, but in the wild may occur in the burrow. The gestation period is about 21 days and the young are born between March and May. In captivity, however, bilbies breed throughout the year. Pouch life lasts about 75 days. Captive bilbies have been observed to take young that had recently left the pouch and deposit them in burrows for about two weeks, returning regularly to suckle them during the night. Captives have lived for more than three years.

Some burrows studied appear to be grouped, suggesting that the animals may live in colonies. However, investigations and reports from Aborigines show that pairs of animals and their latest offspring may inhabit each group of burrows. Home ranges may be temporary in location and shift in response to food availability. Bilbies studied in the Northern Territory were found to stay within 100 metres of their burrows, although they used and visited a number of burrows.

The bilby is placid and adapts well to captivity. In comparison with other bandicoots, it is relatively social, and can be kept in groups without serious fighting. Males may develop a dominance hierachy, reinforced by scent marking from glands in the urogenital area. Females have been reported to share burrows with males and other females, but dominant males will nip and chase other males away from burrows. A quiet, open-mouthed hissing, one of few vocalisations recorded, may be used as a threat in such situations.

The bilby once occurred over a wide range in arid and semi-arid regions in mainland Australia south of about 18 degrees latitude. It has long disappeared from the better watered parts of this range. Its former habitat included savannah woodland and it now mainly inhabits arid hummock grassland and acacia shrub land with spinifex or tussock grass. It has also been recorded in *Melaleuca* thickets in

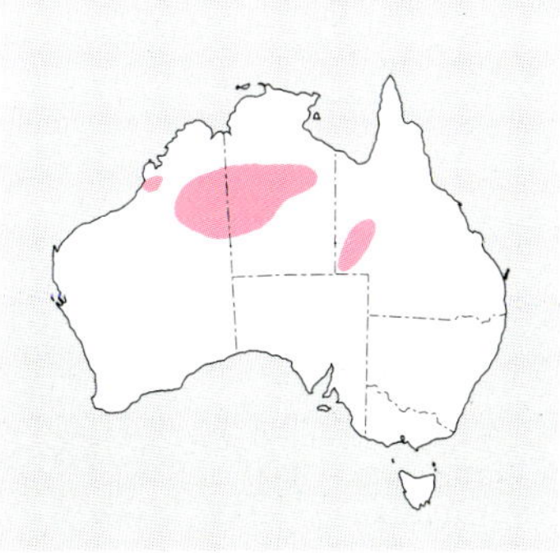

Northern Territory: An area roughly bounded by Renner Springs in the north, the MacDonnell Ranges in the south, Delmore Downs in the east and the Tanami Desert in the west.
Western Australia: Dampier Land, and around the Edgar and Warburton ranges.
Queensland: A small isolated area in the southwest Channel Country between Boulia and Betoota.

The bilby blocks the entrance to its burrow with soil which extends for some distance into the tunnel.

the Northern Territory. Recent studies have shown that it prefers areas with lots of open ground.

It was once widely distributed over the southern parts of the Northern Territory but is now restricted there to scattered colonies in an area roughly bounded by Renner Springs in the north, the MacDonnell Ranges in the south, Delmore Downs in the east and the Tanami Desert in the west. It was also once widely distributed over most of Western Australia south of the tropics, but is now restricted to small populations in Dampier Land and around the Edgar and Warburton Ranges. In Queensland the bilby is presently restricted to a small isolated area in the southwest Channel Country between Boulia and Betoota. Although it occurred widely in New South Wales as far east as the western slopes of the Great Dividing Range, it is almost certainly extinct there now for the last record was in 1912. It was also formerly known, too, in northwestern Victoria and north of the mouth of the Murray in South Australia. The last record for Victoria is 1896 and although they may have persisted in the Lake Eyre Basin, South Australia, until the 1960s, there have been no subsequent recordings in that area. There is evidence of a continuing decline in its remaining range. A number of subspecies are already thought to be extinct: *Macrotis lagotis cambrica* (New South Wales), *Macrotis lagotis nigripes* (South Australia), *Macrotis lagotis grandis* (South Australia), and *Macrotis lagotis interjecta* (southern Western Australia).

The sudden and widespread contraction of the bilby's range may be attributable to the effects of rabbit and livestock grazing, foxes and cats and a change in the fire regime implemented by Aborigines. This changed the type, adundance and diversity of food plants. It may be necessary to reintroduce Aboriginal fire regimes to ensure survival in remaining habitats. A World Wildlife Fund Australia study aims to determine distribution and abundance, and assess future conservation requirements.

It must be remembered that, even though it is much reduced in numbers the bilby remains one of the least affected of central Australia's middle-sized mammals. It is to be hoped that in the future we may say, as Baldwin Spencer said during the Horn Expedition to central Australia in 1894, that its 'burrows abound'; and that we may return to the situation that Finlayson noted when he described the bilby as 'one of the most plentiful and universally distributed of central Australian mammals'.

BRIDLED NAIL-TAIL WALLABY

(Onychogalea fraenata)

FROM
The Mammals of Australia,
John Gould, 1848-1863.
Gould's name for the mammal:
Bridled Nail-tailed Kangaroo,
Onychogalea fraenata.

Thanks to a report on extinct animals in the periodical Women's Day *the bridled nail-tailed wallaby was rediscovered.*

The medium-sized bridled nail-tail wallaby has one of the most distinctive colour patterns of any kangaroo species. First described scientifically by John Gould in 1840, this lightly built animal is about the size of a large hare. It weighs 4-6 kg and has a head and body length of 430-700 mm, with the tail adding another 360-540 mm. There is some variation in size between the sexes. The popular name of this pretty animal refers to the markings found on the head and shoulders, which are suggestive of a bridle. These consist of a white line that runs from the centre of the neck down behind the forearm on each side of the body, and a less distinct white cheek-stripe. There is a distinct black mark in front of the white bridle. The general colour of the soft, thick fur is clear grey, while the underfur is a dark slate colour at the base, grading to light grey at the tip. The animal has a hairy, narrow nose and a black back stripe that widens on the neck. The fingers and extreme tip of the tail are black and there is a horny pointed 'nail' on the tip of the tail.

Information about the biology and behaviour of this species remains limited. The female's pouch contains four teats, and a single young is produced, usually in May. Little has been recorded of the social behaviour of the species, although Gerard Krefft described it as 'gregarious'. Other writers have mentioned that it is usually solitary and appears in larger numbers only when the dry season progresses and the pasture deteriorates. Females may be seen with young at heel, but larger groups of four or five animals have also been observed. Krefft, writing in 1866, noted that although it was often seen during the day, specimens which he observed in captivity were more active at night. More recent observations show that this animal rests in slightly dished patches of bare ground, scratched out beside a bush or tree, during the day and that it commences feeding around dusk. It feeds on grasses, herbs and shrubs and seems to prefer the transitional zone between brigalow scrub and open eucalypt forest, where there is an open canopy and grassy understory. There have been few sightings recorded in cleared areas.

The bridled nail-tail wallaby was known as 'Flash Jack' in the Wagga Wagga area, which was part of its range in the nineteenth century. It is a very apt name for this handsome, swift wallaby which hops with a fast, smooth movement and with its body held low to the ground. It may hide in long grass in a prone position to avoid detection by predators, and will remain hidden until closely

Vulnerable	*Could become endangered because few in numbers or limited in distribution.*
Endangered	*Likely to become extinct unless action is taken.*
Critical	*Only a few hundred individuals survive, or only a single small population known.*
Extinct	*Not sighted for more than 50 years.*

BRIDLED NAIL-TAIL WALLABY

Onychogalea fraenata

Head and body length: 430-700 mm

Tail length: 360-540 mm

The general colour of this mammal is clear grey, but if you look at an individual hair, it is dark slate at the base, changing to light grey at the tip. Bridled nail-tail wallabies often feed upon fallen leaves in brigalow scrub in the small area of central Queensland where they are still found. Their swift movements and ability to elude pursuers have earned them the nickname of 'Flash Jack'.

(H. & J. Beste/NPIAW)

By the 1960s the species was presumed extinct. It was not until 1973 that the bridled nail-tail wallaby reappeared.

Queensland: Near Dingo.

approached. Gould related its ability to run fast and hide in tree hollows when chased: '. . . it bounds away with remarkable fleetness, generally giving the best dogs a sharp run, and frequently making its escape by gaining the thick part of the bush, or the hole of a decayed tree; one of those procured, on being sharply pressed, mounted the inside of the tree, to an opening nearly fifteen feet from the ground; whence it leaped down before the dogs and succeeded in reaching the hollow of a fallen trunk, from which it was finally taken by hand' (Gould 1863). Those enduring naturalists of the nineteenth century, Gould and Krefft, shared the Aborigines' appreciation of the animal's fine culinary attributes. Gould thought the flesh was like that of other small kangaroos — excellent, and to be preferred over other types of meat. In one of the only two sentences that he devoted to the species, Krefft tell us that its flesh was white and 'well tasted'.

The bridled nail-tail wallaby was once plentiful over much of inland New South Wales west of the Great Dividing Range, and ranged from the Murray River, Victoria, in the south, to Charters Towers Queensland, in the north. Today its only know remaining population is a small one near Dingo in central Queensland. The decline of the bridled nail-tail wallaby has been rapid. In 1866, when writing about northwestern Victoria and southwestern New South Wales, Krefft called it 'the most common of all the smaller species of the kangaroo tribe'. At the turn of the century it was still sufficiently common for bounties to be paid for its scalp. However, the animal became extremely rare soon after this time. The last record of its existence in New South Wales is from an animal taken near Manilla, in the northeast of the state, in 1924. It seems to have disappeared from Victoria long before this. By the 1960s the species was presumed extinct, as the last authenticated record of its existence was from the Dawson Valley, Queensland, in 1930.

It was not until 1973 that the bridled nail-tail wallaby reappeared — and then only as a result of a tenuous and lucky connection. Thanks to a report on extinct animals in the periodical *Women's Day,* a Mr Challacombe, fencing contractor of Duaringa, Queensland, recognised a wallaby that he had seen while working near Dingo. As a result of his discovery the Queensland government set up a faunal refuge to help protect the tiny colony that had survived. The government purchased two grazing properties with a total area of 11 240 hectares, thus securing 95 per cent of the remaining habitat

Krefft reported the bridled nail-tail wallaby to be the most common wallaby in country such as this, near the junction of the Murray and Darling Rivers, in the 1850s. It has been extinct for nearly a century in that area, and survives today only in a small area in central Queensland.
(A. H. Burbridge/ANT Photo Library)

of this remnant population. There is some hope that at a future date the animal may be reintroduced to various parts of its former range.

The spectacular decline and subsequent re-emergence of the bridled nail-tail wallaby is a story to gladden the hearts of environmentalists. Unfortunately most other species that have been presumed extinct have not been so lucky. Just why this species should have vanished from areas that had not been extensively cleared for timber or overrun with introduced predators remains a mystery. Factors contributing to its decline may have been the destruction of ground cover, competition for food from introduced stock animals and the introduction of the fox. It is to be hoped that the species will remain common in its limited range.

It may hide in long grass in a prone position to avoid detection by predators.

It is able to use its tail, curled around in a prehensile manner, to carry bundles of nesting material.

Vulnerable	*Could become endangered because few in numbers or limited in distribution.*
Endangered	*Likely to become extinct unless action is taken.*
Critical	*Only a few hundred individuals survive, or only a single small population known.*
Extinct	*Not sighted for more than 50 years.*

The brush-tailed bettong was once very common over much of southern Australia. There has been some disagreement about the number of sub-species that made up populations in various areas and the matter has been complicated by the lack of specimens from some areas. As a result it is very difficult to make comparisons. Two sub-species are recognised at present: *Bettongia penicillata penicillata* (southeastern Australia), and *Bettongia penicillata ogilbyi* (southern and southwestern Australia). *Bettongia tropica* from north Queensland has sometines been considered a sub-species of *Bettongia penicillata.* However, evidence supporting this view is scant.

The brush-tailed bettong was described scientifically by Gray in 1837. Measurements of the only remaining population (in south-western Australia) are: head and body length 300-380 mm; tail length 290-360 mm. The animal weighs 1.1-1.6 kg, with the weight diminishing significantly in winter. The overall colour is brownish grey. The sides of the short face and the bases of the ears have a reddish tinge, while the undersides are pale yellowish grey. The base of the long tail is also brownish grey and the last two-thirds is dark brown to black above. The tail is crested towards the tip.

The brush-tailed bettong breeds prolifically throughout the year. In common with many other kangaroos, it comes into heat shortly after giving birth and can keep embryos in a kind of suspended animation (called embryonic diapause) until they are needed. Females can produce young when only six months old. A single young spends 90 days in the pouch and there is an interval of about 100 days between each birth. The weaned young share the nest and accompany the mother at heel until they are displaced when a new infant leaves the pouch.

Studies of the remaining population have shown that this animal consumes neither green material or water. Indeed, it has a most unusual diet for a mammal. While it may eat bulbs, tubers, seeds, insects and resin, the bulk of its nutrients (especially in summer and autumn) are derived from the fruiting bodies of underground fungi. These fungi can only be digested indirectly. Bacteria in the large forestomach consume the fungi and it is these bacteria and their by-products which contain the more balanced nutrients that are digested in the rest of the stomach and small intestine. The strong foreclaws are used for digging food. The animal has also been seen burying seeds but it is not known if it ever recovers them. However the seeds have been

BRUSH-TAILED BETTONG

(Bettongia penicillata)

FROM
The Mammals of Australia,
John Gould, 1848-1863.
Gould's name for the mammal:
Jerboa Kangaroo,
Bettongia penicillata.

BRUSH-TAILED BETTONG

Bettongia penicillata

Head and body length: 300-380 mm
Tail length: 290-360 mm

The brush-tailed bettong's short face has grizzled brownish-grey sides, and the bases of the ears have a reddish tinge. The undersides of the body are pale yellowish grey. The tail has a crested tip.

(Jean-Paul Ferrero/AUSCAPE)

It has a most unusual diet for a mammal, eating fungi that can only be digested indirectly.

known to germinate in clumps after fires.

In common with some other potoroines, the brush-tailed bettong has another curious habit, which was commented upon by John Gould. It is able to use its tail, curled around in a prehensile manner, to carry bundles of nesting material. Gould was amused by this feat '. . . their appearance when leaping towards their nests with their tails loaded with grasses is exceedingly grotesque and amusing . . .' (Gould 1863). The brush-tailed bettong is also strictly nocturnal and is not gregarious. During the day the animal rests in a well-made and hidden nest which consists of grass and shredded bark. A shallow depression is dug under a bush or similar cover. It is comparatively slow-moving but hops rapidly when disturbed or frightened, with the head held low, the back arched and the tail extended almost horizontally. It is one of the few mammals encountered by the nineteenth-century naturalist Gerard Krefft whose culinary qualities do not rate a mention!

The remaining habitat of this species is characterised by open eucalypt forest with an understorey of tussock grass, low woody scrub and occasional bare patches of ground. Gould related a similar habitat for the former population in eastern Australia: 'The usual resorts of the Jerboa Kangaroo are low grassy hills and dry ridges, thinly intersected with trees and bushes . . .' (Gould 1863). The species appears to be particularly adapted to habitats that are subject to frequent fires. As a result, the halting of Aboriginal fire regimes may have had a great bearing on its dramatic decline. It is said that it can survive fires by racing ahead of the blaze or backtracking into the already burnt bush. Its food and shelter plants are also well adapted to fire.

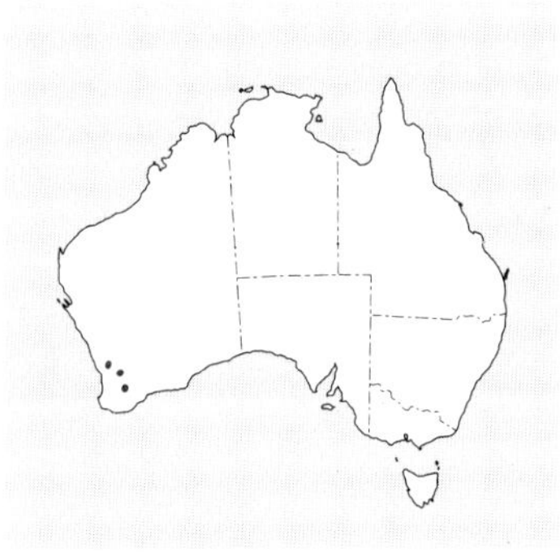

Western Australia: The Perup and Dryandra forests and the Tuttanning Reserve.

At the time of European colonisation the species was known to occur from the west coast, across southern Australia to north-western Victoria, and in inland New South Wales to the western slopes of the Great Dividing Range. Gould did not report its existence between the New South Wales coast and the Great Dividing Range, although he noted its abundance on the Liverpool Plains and other inland areas. The naturalist and explorer Hedley Finlayson gave the northern limit of distribution as about 20 degrees south latitude. The species also formerly existed on St Francis Island, Nuyts Archipelago, South Australia, but is thought to have disappeared from there by the early 1900s as a result of farming and the introduction of cats. Professor F. W. Jones, writing in 1923, noted the tragedy

Woodland such as this in the southwest of Western Australia provides the last habitat of the brushtailed bettong.
(Frédy Mercay/ANT Photo Library)

of the species' disappearance from South Australia: 'Twenty years ago the dealers in Adelaide did a great trade in selling them by the dozen at about ninepence a head for coursing on Sunday afternoons'. Despite the former abundance of the species, not a single South Australian specimen was preserved for scientific study. The species survives only in three small areas of Western Australia: the Perup and Dryandra Forests and the Tuttanning Reserve. Studies have been made and continue to be conducted in these areas, and the habitat is being managed to help ensure the continuing survival of the species. Captive breeding colonies have also been set up in an attempt to introduce the species to several South Australian islands.

The brush-tailed bettong's decline seems to have been caused by a number of factors. These include a change in fire-frequency, agricultural clearing, competition and disturbance from stock and predation by humans and foxes. It is hoped that, given time, the species may recover some of its former range and no longer be endangered.

BURROWING BETTONG

(Bettongia lesueur)

FROM
The Mammals of Australia,
John Gould, 1848-1863.
Gould's name for the mammal:
Gray's Jerboa Kangaroo,
Bettongia graii.

It uses scent to locate tubers and bulbs, which it then digs out of the ground.

This stocky marsupial is the only member of the kangaroo super-family that regularly digs and inhabits burrows all year round. A number of distinct forms have been described in the past, but it is now thought that these are simply variations within a single species. The burrowing bettong is yellowish grey-brown with a lightly haired, fattened tail which has virtually no tail crest. In some regions, the tail has a white tip. A pale, indistinct hip-stripe is usually present. A curious aspect of the burrowing bettong's yellow colouration has been mentioned by Professor F. W. Jones. In museum skin specimens the yellow colouration of the pelt fades quickly and can even be removed by the action of alcohol, used in the preservation of specimens. The moderately haired ears are shorter and more rounded than in other bettong species. The burrowing bettong exhibits variability in size, weight and relative tail length throughout its range. The head and body length is 280-400 mm, with the tail adding another 215-300 mm. It weighs around 2 kg, but can be lighter in some areas. The southwestern variety is larger, while animals from Barrow Island are significantly smaller.

The first specimen of this animal was obtained in 1817 from Dirk Hartog Island by the famous navigator de Freycinet during his expedition in *L'Uranie*. The animal is now extinct on this island.

The diet of this strictly nocturnal animal is largely vegetarian. It uses scent to locate its food, which it then digs out of the ground. It consumes tubers, bulbs, seed nuts and the green parts of plants. One of the seeds it eats is that of the quandong, *Santalum acuminatum*. Communities on Barrow Island are known to eat native fig fruits, seeds, roots, fungi and termites. Troughton, writing in 1967, noted that the animals on the Shark Bay islands fed mainly on marine refuse, '. . . even dead sheep were occasionally partly eaten'. John Gould's collector, John Gilbert, related how this once abundant animal raided settlers' gardens in Western Australia: '. . . it is one of the most destructive animals to the garden of the settler . . . almost every kind of vegetable being attacked by it, but especially peas and beans . . .' (Gould 1863).

The burrowing bettong can breed throughout the year and may give birth to and rear as many as three successive young in a year. The estrous cycle of the female is a comparatively short 23 days, while gestation takes approximately 21 days. Like many other kangaroo species, the burrowing bettong has 'embryonic diapause'. This allows another young to be born 20 days after a juvenile is lost from the

Vulnerable	*Could become endangered because few in numbers or limited in distribution.*
Endangered	*Likely to become extinct unless action is taken.*
Critical	*Only a few hundred individuals survive, or only a single small population known.*
Extinct	*Not sighted for more than 50 years.*

BURROWING BETTONG

Bettongia lesueur

Head and body length: 280-400 mm

Tail length: 215-300 mm

The presence of a white tail tip is variable in this species. The animal is yellowish-brown with an indistinct hip stripe and its tail is quite fat. The rabbit, the fox and the use of poison grain have all been blamed for causing its decline.

(Jean-Paul Ferrero/AUSCAPE)

It is very vocal and can make a variety of squeals, hisses and grunts.

pouch. The young leaves the pouch at about 115 days after birth but is not weaned for a further three to ten weeks. Sexual maturity is attained within a year.

Professor F. W. Jones noted the pugnacious nature of the species, especially the males. He related how males will fight to the death if kept together in captivity, scratching and plucking fur. However, this aggressive nature seems inconsistent with the animal's communal way of life; it often shares burrows with other burrowing bettongs and even with rabbits. Jones' animals may have been under stress, or perhaps strangers to each other, although Gilbert noted that nearly every specimen he saw was denuded of fur on the back in a fashion similar to a dog's mange. This may have been a result of fighting. The burrowing bettong has a good sense of smell and vision, and hearing appears to be well developed. It is very vocal and can make a variety of squeals, hisses and grunts. The animal moves in a bipedal fashion and, except when it is stationary, does not use its tail and front limbs for support.

The burrow varies considerably according to circumstances. It may be a simple structure with only one or two entrances and a short, curving tunnel, or a large warren with more than 100 entrances. One of these warrens may house more than 50 individuals. Both Jones and Gerard Krefft noted the great difficulty of extracting animals from such warrens. Nests made of vegetation can be found within the burrow system. Like other bettong species, the burrowing bettong can use its tail to carry nesting material. Warrens were constructed in many types of soil on the mainland. However, loams were favoured and in sandridge deserts the burrows were situated in the damper low-lying areas between dunes. Slight outcrops of limestone or gypseous rock and rises in salt lake systems were other favoured habitats. In rocky areas burrows may be situated under boulders or capping rock. On the mainland, habitats ranged from open eucalypt or acacia woodland with a grass and shrub understorey to sandridge desert clothed with spinfex hummocks and sparse shrubs. In island habitats the vegetation cover consists of a variety of shrubs, herbs and grasses.

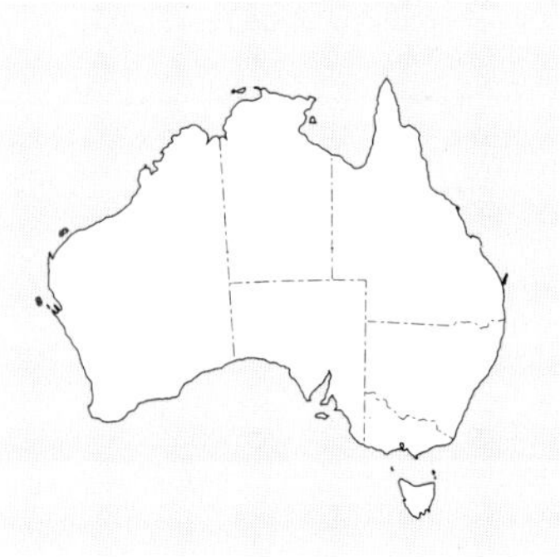

Western Australia: Bernier and Dorre islands in Shark Bay; Barrow Island and nearby Boodie Island.

The burrowing bettong's range has declined greatly since European settlement. It was once common over a large part of Australia, from west of the Great Dividing Range to coastal Western Australia. Tasmania is the only state from which it was certainly absent. Thus it had one of the greatest geographical ranges of any Australian mammal. Today it is confined to four islands off the west coast:

The burrows of burrowing bettongs were once abundant throughout central Australia. But its destructive habits made it a nuisance to settlers for it attacked their crops and vegetable gardens. In areas suitable to the burrowing bettong it honeycombed the land with its burrows, which had more entrances than there were occupants.
(K. Johnson/NPIAW)

Bernier and Dorre Islands in Shark Bay, and Barrow Island and nearby Boodie Island of northwestern Australia. It is presumed extinct throughout the mainland and on Dirk Hartog Island, whence the first specimen was obtained. It had disappeared from Victoria and New South Wales by the 1860s and from southwestern Australia by the 1930s. It persisted, however, at least until the 1940s in some central desert areas. Krefft, writing in 1866, related how the animals were retreating before the herds of cattle. By that time it was no longer to be found south of the Murray, and Krefft did not obtain a single specimen in Victoria. These animals were once so abundant in South Australia that Jones related how they were a constant annoyance to rabbiters, who continually found them sharing warrens with rabbits. Jones suggested that rabbits, foxes and the use of poison grain were the cause of the species' decline. Fortunately, it seems to be secure on its four island sanctuaries.

DIBBLER

(Parantechinus apicalis)

FROM
The Mammals of Australia,
John Gould, 1848-1863.
Gould's name for the mammal:
Freckled Antechinus,
Antechinus apicalis.

It has a jerky way of moving and quickly seeks cover if it is disturbed.

The dibbler was discovered by Europeans as early as 1838, and was described by 1842. Despite its early discovery it has proved to be an elusive species. A member of the marsupial family Dasyuridae, it was long placed in the genus *Antechinus* before being re-classified. In many books it will still be found listed under its earlier name.

The head and body length of the dibbler has been measured as 145 mm for males and 140 mm for females. The tail has been measured as 105-115 mm for males and 95 mm for females. Weight varies between 40 and 100 grams, depending upon age, sex and habitat. It has coarse, grey-brown fur flecked with white tips on its upper surface, and long, soft, thick greyish-white fur tinged with yellow underneath. It has a tapering hairy tail and a distinct white ring around the eyes.

The dibbler breeds in autumn, the females being receptive to the advances of males for only a few consecutive days. The male produces sperm between the months of January and April. The gestation period is thought to be between 44 and 53 days. The female has eight nipples in a pouch area that only develops fully while the young are being carried. There can be as many as eight young in a litter. The young remain dependent upon their mother for around four months and are ready to breed themselves when only 10 to 11 months old. Single copulations may last for several hours, but they do not, as in some other dasyurid species, result in the exhaustion and death of the male at the end of the breeding season. In fact there may be more than one copulation per breeding season, and individuals may survive for two to three years, breeding during each season. Although seven pouch young have been reared to maturity in captivity, captive dibblers have never bred successfully. The dibbler lives on coastal sandplains in areas where waist-high heathland thickets containing many stunted banksias have developed. It has a jerky way of moving and quickly seeks cover if it disturbed. It scampers through the leaf litter and has been seen climbing branches and twigs of banksias in search of insects and nectar. The habits of the recently discovered colony on Boulanger Island are, however, markedly different. Here there are no trees to climb, and they apparently spend around 80 per cent of their foraging time in seabird burrows excavated among the roots of sand dune shrubs. Dibblers are known to eat chopped meat and insects in captivity, and studies of faeces of Boulanger Island dibblers indicate that they feed upon a wide variety of invertebrates,

Vulnerable	*Could become endangered because few in numbers or limited in distribution.*
Endangered	*Likely to become extinct unless action is taken.*
Critical	*Only a few hundred individuals survive, or only a single small population known.*
Extinct	*Not sighted for more than 50 years.*

DIBBLER

Parantechinus apicalis

Head and body length: 140-145 mm
Tail length: 95-115 mm

The dibbler has coarse, grey-brown fur flecked with white tips on its back, with long, soft, thick greyish-white fur tinged with yellow underneath. Insects such as this mole cricket form part of the diet of this rare marsupial.

(D. Whitford/NPIAW)

It has been seen climbing branches and twigs of banksias in search of insects and nectar.

small birds, reptiles and occasionally house mice *(Mus musculus)*, as well as a considerable amount of green plant parts. The dibbler is thought to be largely nocturnal, although captive animals emerge during the day to bask. Accounts by Aborigines indicate that dibbler's nests consist of small depressions covered with mounds of grass and twigs. However, no nests have yet been discovered by biologists.

As a result of some old and confused records, the dibbler was once thought to occur as far afield as South Australia and Rockhampton in Queensland. It now seems that these records were erroneous, and that the dibbler was always restricted to the southern and western parts of Western Australia. During the early days of European settlement the dibbler was recorded from areas of the southwest, such as New Norcia and Moore River, where it no longer exists. In addition, fossils, sometimes in large numbers, have been found as far north as Shark Bay. Today however the dibbler is extremely rare on the mainland, being restricted to a few coastal localities in the extreme southwest.

The reason for this decline is not clear, however a change in fire regime that followed European colonisation may have been a major factor. Indeed, so precipitous was its decline that the dibbler was long thought to be extinct; no sightings or records of the species were made for 83 years between 1884 and 1967. In the latter year however a few specimens were trapped in banksia heathland at Cheyne Beach (now known as Hassell Beach), some 50 km east of Albany. By 1982 numerous intensive surveys in the Cheyne Beach area had located only seven individuals, and the last of those were found in 1976. Some of these animals were retained to start a breeding colony which unfortunately failed. A second occurrence came to the attention of scientists when two dead male dibblers were brought in by domestic pets near Jeracuttup, 225 km northeast of Cheyne Beach. The small area of banksia-dominated heathland where they were found is now a 75 hectare fauna and flora reserve. A third mainland occurrence was discovered in 1985 in the Fitzgerald River National Park.

In December 1985 two further remarkable colonies of dibblers were discovered by zoologists working for the Western Australian Department of Conservation and Land Management. These were on two tiny islands located around about two kilometres offshore in Jurien Bay. Boulanger Island is 25.9 hectares in extent and Whitlock Island only about eight hectares. Trapping on Boulanger Island has shown that dibblers are present all over the island and that there

Coastal heaths, such as this near Denmark in the southwest of Western Australia, still shelter relic colonies of the elusive dibbler.
(Frank Park/ANT Photo Library)

Western Australia: A few coastal localities in the extreme southwest; near Albany; Jeracuttup; the Fitzgerald River National Park; Boulanger Island and Whitlock Island in Jurien Bay.

are equal numbers of males and females. Only a small proportion of females (0.2 per cent) had previously bred. The total population of both islands has been estimated as only around 300. The dibblers of Boulanger Island are presently being studied. Both islands need to be kept free of introduced pests if the dibblers' survival is to be ensured. For the existing mainland colonies the establishment of a fire management program seems to be essential if their precarious existence is to be maintained.

DUSKY HOPPING-MOUSE

(Notomys fuscus)

Sand dune country such as this in inland Australia still supports small colonies of the dusky hopping-mouse. It forages amongst cane grass and stunted acacias.

(Otto Rogge/ANT Photo Library)

Their burrows have pop-holes that are about three centimetres in diameter, and shafts that extend downwards for 1.5 metres.

The delicate-looking dusky hopping-mouse, with its attractively coloured silky fur, large ears and dark, prominent eyes, is perhaps the most beautiful of all Australian rodents. It was the great naturalist Professor F. W. Jones who first recognised it as a distinct species in 1925, giving it the name *Ascopharynx fuscus*. The fur is light orange to grey on the back and white on the underside. Both males and females have a distinctive throat pouch, with well-defined fleshy margins, which is covered by stiff white hairs that point towards the centre of the pouch. The dusky hopping-mouse shares many external and skeletal similarities with some other species of hopping-mice within its range. In fact it was once thought to be merely a variety of the fawn hopping-mouse, *Notomys cervinus*, which occurs in the same area. However the dusky hopping-mouse has a better developed throat pouch than other species. As well, it has distinctive chromosomes as well as some other structural differences.

The head and body length is 80-115 mm, while the tail, which is longer than the body, measures 115-155 mm. The animal can weigh anything between 30 and 50 grams. The dusky hopping-mouse has short, brown hairs on the upper surface of its tail and white hairs below. It has a tuft of hairs at the end which is mainly brown with a small amount of white underneath.

There is little information available on the breeding habits of the dusky hopping-mouse in the wild. A small number of pregnant females have been caught in both summer and winter. Young have also been recorded in April and the winter months. The animal breeds throughout the year in captivity. The oestrous cycle is eight days with the gestation period lasting 38-41 days. Females return to oestrous immediately after giving birth. The most common litter size is three although it can vary between one and five. The eyes of the young open at about 20 days and they are weaned at about 30 days. They become sexually mature when they are between 75 and 90 days old.

The diet of the dusky hopping-mouse consists mainly of seed with some green plant material and insects. Sufficient water is obtained through this diet. Little is known of the habits of this species in the wild. Studies suggest that colonies may have an attachment to particular localities on sand dunes; the animals can sometimes be found in a loose group on one sandhill but not at all on adjacent hills. The dusky hopping-mouse digs its burrows on the flat top of a dune. This burrow is deep and simple and has a number of entrances.

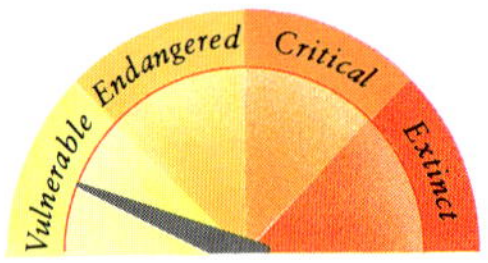

Vulnerable	*Could become endangered because few in numbers or limited in distribution.*
Endangered	*Likely to become extinct unless action is taken.*
Critical	*Only a few hundred individuals survive, or only a single small population known.*
Extinct	*Not sighted for more than 50 years.*

DUSKY HOPPING-MOUSE

Notomys fuscus

Head and body length: 80-115 mm

Tail length: 115-155 mm

The colouring of this species ranges from light orange to grey on the back, with white on the underside. Its large eyes and ears are adaptations to a nocturnal existence.

(H. & J. Beste/NPIAW)

The dusky hopping-mouse climbs shafts in its burrow by hopping, but descends with a headlong dive.

The pop-holes are about three centimetres in diameter and shafts may extend downwards for 1.5 metres. The shafts are interconnected by horizontal tunnels and contain a nest chamber with a pad of chewed and shredded vegetation. The burrow systems appear to be connected on the surface by well-marked runways. This suggest that burrow systems may be used by more than one group of mice.

The dusky hopping-mouse lives in groups of about five individuals and may use more than one burrow system. The animal does not usually venture onto the gibber plains, as does the fawn hopping-mouse, but prefers to forage for food during the night on the top and sides of sand-dunes. The day is spent in the burrow. The animal climbs the shafts by hopping but descends with a headlong dive. If the burrow is blocked with sand, the animal will squeak several times to summon other mice within the burrow to help clear the blockage.

In southwestern Queensland the animal's habitat is the sandridges which alternate with gibber flats and claypans. Cane-grass and stunted acacias are typical vegetation. The dusky hopping-mouse once lived in an area of inland Australia that stretched from Ooldea in South Australia to southwestern Queensland. One record is from as far afield as Rawlinna in southeastern Western Australia. The population density is low but may once have been greater. The explorer Charles Sturt reported upon the abundance of a species of hopping-mouse that was probably the dusky hopping-mouse. During a journey to central Australia in 1845 Sturt met a group of Aborigines near the northern edge of the Flinders Ranges: '. . . These wanderers of the desert had their bags full of jerboas which they had captured on the hills. They could not indeed have had less than from 150-200 of these beautiful little animals, so numerous are they on the sand hills . . . our friends cooked all they had in hot sand, and devoured them entire, fur, skin, entrails and all, only breaking away the under jaw and nipping off the tail with their teeth . . .' (Sturt 1849).

As the animal may have diminished in population density and certainly done so in range, it remains vulnerable to external changes to its habitats and must be considered endangered. The last word must go to its discoverer, the early conservationist Professor F. W. Jones, who, writing about the dusky hopping-mouse in 1925, made the following observations. 'It is easy to see that without extensive collecting, careful preservation, and patient study we shall not attain to any real knowledge of our smaller mammals before the day of

Queensland: Southwestern sandridges which alternate with gibber flats and claypans.

their individual extermination arrives. In Australia all too little has been done, and although so much remains to do and so little time will be available for its accomplishment, there are as yet few indications that a public spirit or administrative sympathy are developing to encourage the endeavour . . .' We can only hope that history will eventually prove him wrong.

GHOST BAT

(Macroderma gigas)

The ghost bat inhabits limestone caves in areas like this in the Kimberleys. It is found only in Australia and is the only representative of its family (false vampire bats) to be found here.

(Adrian G. Davey/ANT Photo Library)

The ghost bat is named for its appearance, created by its whitish flight membranes and pale grey back fur.

The ghost bat is found only in Australia, and it is the only representative of its family (false vampire bats) to be found here. It was described in 1880 by Dobson. The original specimen, which was sent to the Gottingen Museum, has been lost; it was, presumably, destroyed by bombing during World War II. The animal's large size and carnivorous nature have caused it to be confused with the true South American vampires, and it has been erroneously described as the 'great blood-sucking bat'. The ghost bat could more accurately be described as a predator upon its neighbours, as part of its diet consists of the smaller insectivorous bats that share its roosts.

The ghost bat is named for its pale appearance, created by its whitish flight membranes and pale grey back fur. This comparatively large bat is 100-130 mm long and has a forearm length of 102-112 mm. Its weight varies between 140 and 165 grams. Its very large ears, which stand directly above the face, are rounded and joined in the middle for more than half their length. It possesses large eyes, a fairly simple spear-shaped noseleaf, and has no externally visible tail.

The ghost bat mates sometime in autumn or winter, and the female produces a single young between September and November. At first the baby is carried by its mother. Then nursery colonies are formed in which the mothers feed prey to the weaned juveniles. The males remain apart from these colonies. When, at seven weeks old, the young can fly, they accompany their mothers to forage until they are fully independent. They are weaned within a few months. Information on the longevity of ghost bats is scant, although two of them are reported to have been kept in captivity at Tubingen University for more than 16 years.

The ghost bat is Australia's only carnivorous bat. Its diet consists of other bats, rodents, and other small mammals, large insects, frogs, lizards and birds. Unlike some other bats, the ghost bat does not make extensive use of echolocation in locating and catching its prey. Instead, it appears to use its eyes and to listen for sounds made by its prey as it perches from a vantage point. During feeding activity ghost bats are known to produce loud bird-like chirps. Much of their prey is captured on the ground or on a vertical surface. After locating a potential meal the ghost bat flies over the animal, and it is during this flight and just before take-off that echolocation pulses are emitted. The attack is very brief, lasting only a second. The prey is killed by a bite to the head or neck region. The victim is then taken to

Vulnerable — *Could become endangered because few in numbers or limited in distribution.*

Endangered — *Likely to become extinct unless action is taken.*

Critical — *Only a few hundred individuals survive, or only a single small population known.*

Extinct — *Not sighted for more than 50 years.*

GHOST BAT

Macroderma gigas

Head and body length: 100-130 mm

Forearm length: 102-112 mm

The large eyes, ears, and nose leaf make the ghost bat a highly efficient predator on small vertebrates.

(G. B. Baker/NPIAW)

Limestone is not in short supply. Yet caves that harbour one of the largest known colonies of ghost bats are being destroyed.

a feeding site, such as a rock overhang or small cave, to be consumed. These feeding sites are characterised by large accumulations of discarded prey parts and bat guano.

During daylight hours the ghost bat roosts in caves, mines and rock fissures. Studies suggest that it may roost in areas of 'twilight' rather than in complete darkness. Its secretive nature, however, means that its presence is often overlooked. It does not usually congregate in large numbers, but more often roosts singly or in small numbers. Some of the larger roosting sites, however, are known to contain up to 500 individuals.

In the past 10 000 years the ghost bat has experienced a great reduction in its distribution, particularly in the south and centre of its range. Professor F. W. Jones and subsequent scientists have suggested that this reduction has been caused by the spread of desert conditions within Australia during this time. Increasingly arid conditions would mean that a carnivore would have less prey. Unfortunately for this theory, recent studies have shown that the Australian climate has not become markedly drier during this period; it may indeed have become more moist. The ghost bat has managed to survive in quite diverse habitats — some of them, such as the Pilbara region of Western Australia, quite arid. Jones noted the finding of very well preserved mummies in caves in the Flinders Ranges, and sub-fossil and fossil remains have been found elsewhere in southwestern Australia as well as in the South Australian ranges and eastern New South Wales. The ghost bat's present distribution is patchy, although it is widespread in northern Australia. Its numbers are small enough for it to be considered endangered. As it is sensitive to disturbance, it is important that areas where it is known to occur in any numbers are protected.

It is tragic, especially in these times of increased sensitivity to environmental destruction, that the continued viability of such a unique bat has been threatened by limestone mining at Mount Etna in Queensland. The destruction of caves that harbour one of the largest known colonies of ghost bats is particularly galling when we realise that limestone is not in short supply and can readily be obtained from less environmentally sensitive areas.

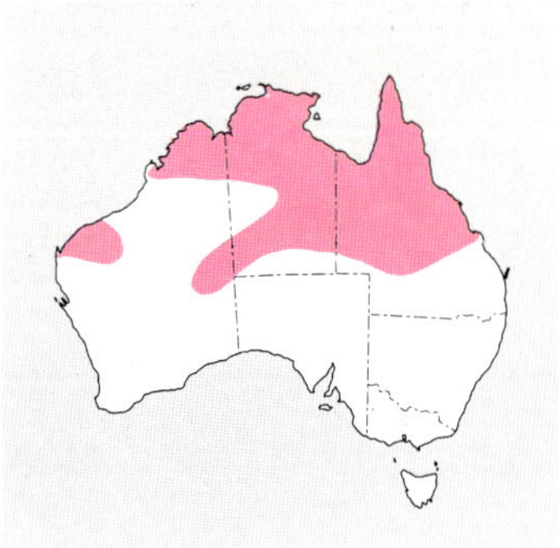

Widespread in northern Australia, but its numbers are small enough for it to be considered endangered.

GREATER STICK-NEST RAT

(Leporillus conditor)

FROM
The Mammals of Australia,
John Gould, 1848-1863.
Gould's name for the mammal:
Building Hapalotis,
Hapalotis conditor.

The babies cling tightly to the mother's teats and are dragged around in this way until they are quite large.

Vulnerable — *Could become endangered because few in numbers or limited in distribution.*

Endangered — *Likely to become extinct unless action is taken.*

Critical — *Only a few hundred individuals survive, or only a single small population known.*

Extinct — *Not sighted for more than 50 years.*

Of the two species of stick-nest rat one is almost certainly extinct and the other, the greater stick-nest rat, is now confined to one small island in the Great Australian Bight. The nest-building habits of these animals are intriguing and have resulted in their common name of 'house-building rat'. Early explorers noticed and commented upon their remarkable nests before they knew what had made them.

The greater stick-nest rat is about the size of a small rabbit. It has fluffy, yellowish brown to grey fur on its back, and creamy white fur on its underside. Its tail is grey-brown above and whitish below and has a slight brush of hairs on the end. It has large, rounded greyish brown ears which, along with its large eyes and rather blunt nose make it an attractive animal. The explorer Sir Thomas Mitchell felt that, except for its long tail, it might readily pass for a small rabbit. The head and body length has been recorded as 170-260 mm, and its tail adds another 145-180 mm. It can weigh anything between 180 and 450 grams.

The greater stick-nest rat most probably breeds in autumn and winter, although there is some doubt. A breeding program undertaken in Monarto Open Range Zoo, South Australia, resulted in young being born in February. On Franklin Island animals have been recorded as breeding in March and April and juveniles have been captured in June. In 1922 Le Soeuf recorded mainland individuals as being with young in July. According to the scientist Professor F. W. Jones, the time of breeding depends more upon the nature of the season than upon the actual time of year. Le Soeuf mentioned that the babies cling tightly to the mother's teats and are dragged around in this way until they are quite large.The young are eight months old when they first breed, and can produce a second litter seven and a half weeks after the first. Jones also mentioned the young's habit of clinging tightly to the mother's nipples: '. . . even if the mother is chased or captured the young are still found adhering so tightly that it is difficult to dislodge them' (Jones 1925). For many years the female's method of carrying the young led visitors to Franklin Island to believe, erroneously, that the greater stick-nest rat was a marsupial. The gestation period of six weeks is one of the longest recorded for an Australian rodent. Litter size can vary between one and four and the young are comparatively large and well developed at birth. They become independent by the time they are one month old. Females may breed in successive years and are fairly long-lived. One female

GREATER STICK-NEST RAT

Leporillus conditor

Head and body length: 170-260 mm

This photgraph was taken on Franklin Island, South Australia. The greater stick-nest rat has a sacculated stomach and large intestinal pouch which enables it to digest bulky food.

(H. J. Aslin/NPIAW)

Much has been made of the gentle nature of this animal and its reluctance to bite when handled.

has been recorded as living to five years old.

The greater stick-nest rat is largely herbivorous. The Franklin Island rats have been recorded as eating pigface *(Carpobrotus rossi)*, nitre bush *(Nitraria schoberi)* and other succulents, with a small amount of grass. The mainland rats probably had a similar diet. Digestion of this bulky diet is aided by a sacculated (divided) stomach and a large caecum (intestinal pouch). The animal will accept a variety of foods in captivity and has been shown to have a liking for sow-thistles. Ellis Troughton of the Australian Museum recorded a possible instance of meat eating when he described the dual use of a nest. Hawks were known to nest on top of the stick nest and Troughton found bones littered in some nests, which suggested the possibility that the rat had developed a liking for the young of hawks.

This species of rodent does not appear to be totally nocturnal. Troughton noted that, while carrying out field work near Ooldea on the edge of the Nullabor Plain, he occasionally surprised a stick-nest rat sunbaking on the stick mound. Much has been made of the gentle nature of this animal and its reluctance to bite when handled. While a great deal has been recorded about the animal's nest-building habits, little else is known about its other habits or biology. Jones noted that it refuses to build stick houses if it is provided with a comfortable bed, but can never resist the urge to shift sticks about. This instinct for moving objects even extends to ridiculously small objects such as small leaves, which it will move for no apparent reason. From more recent observations of captive animals it is apparent that they will readily build nests if given space and materials.

In the wild, nests are usually begun with a shallow scrape in the ground beneath a small bush, which later becomes part of the structure. Troughton noted that, in the absence of a supporting bush where there were strong winds, the nests had small stones on top of and amongst the sticks, '. . . much in the way rocks may be placed on a shed roof to keep it down' (Troughton 1941). Sticks, branches and stones are dragged to the site in the animal's jaws. Larger branches are gnawed to a more manageable size. The nest structure may be as high as a metre with a diameter of two metres. One or two grass nests are constructed inside. Several passageways are usually left within the nest while many entrance or exit holes are made around the base. The nests may be added to over long periods by the original builder and susequent inhabitants. The Franklin Island rat nests vary in form; they range from a few sticks to more

The greater stick-nest rat was abundant in inland areas such as this, only a century ago. Franklin Island, South Australia is its only remaining habitat.
(Otto Rogge/ANT Photo Library)

complex structures, similar to those found on the mainland.

The rats live communally, with up to six individuals sharing a nest. The animals may also share their burrows with mutton birds, rabbits and fairy penguins. Although the greater stick-nest rat is known for its social nature, it can be hostile to unfamiliar members of its species. Care needs to be taken when introducing new animals into captive communities, as the newcomers may be killed.

The curious nests of the greater stick-nest rat were first observed by Surveyor-General Mitchell during his expedition of 1838, but it was left to Charles Sturt to associate the right animal with the nests. At first Mitchell thought that the piles of wood were for Aboriginal signal fires; it was only on examining them more closely that he found them to be nests. He found them to be extremely well built

The nests are so firmly woven that they are difficult to tear apart.

The nest structure may be as high as a metre.

and difficult to open. Troughton also noted the toughness of the nest construction: '. . . the nests are so firmly woven that we found dragging them apart a difficult and trying task in the heat and dust' (Troughton 1941). Jones mentioned that even where the animal had been exterminated for a considerable time the massive nests still survived. This makes it difficult to rule out the possibility that the species still exists on the mainland. John Gould's representation of the nest has an unrealistic smoothness which may have resulted from Sturt's description of the nests as being like beehives.

The greater stick-nest rat was once widely distributed, in an area extending from the lower Darling River to the Nullarbor Plain. Subfossils show that it once extended further west into Western Australia. Its northern limit seems to have been latitude 28 degrees south. There seems to be some habitat overlap between the two species of stick-nest rat. The greater stick-nest rat is primarily adapted to arid terrain, preferring semi-arid or arid areas of low shrubland or shrub steppe. It had become rare by the 1850s, and Krefft predicted its extinction when he wrote in 1866 that: '. . . it is one of the many species which will soon be extinct, as I found that it had already retreated before the herds of sheep and cattle across the Murray'. Jones reported its existence on the Nullarbor Plain in the 1920s but it appears to have become extinct on the mainland some time in the 1920s or 1930s.

The greater stick-nest rat is now found only on Franklin Island, where 1500-2000 individuals survive. This colony of stick-nest rats was long considered a separate species but is now thought to be the same as the mainland variety. Its survival on Franklin Island may be attributable to the absence of stock animals which disturb and trample nests, and which may also compete with the rats for food. It has two natural predators of Franklin Island: the barn owl and the black tiger snake. Although its numbers are not disminishing on the island sanctuary, the small size of its habitat makes it potentially vulnerable. The island was declared a conservation park in 1967 and attempts are being made to establish a viable breeding colony that can be introduced to another suitably isolated island.

South Australia: Franklin Island.

LEADBEATER'S POSSUM

(Gymnobelideus leadbeateri)

The leadbeater's possum inhabits mountain ash forest in Victoria. However, very old hollow trees and not just young ones such as in this photo must be present before the area is suitable. Forestry operations which eliminate old trees are hazardous for the survival of the species. There are no records of this possum recolonising a clear felled and regenerating site.

(J. W. Kelly/ANT Photo Library)

Captive females have been observed to use force to disperse their own female offspring.

Leadbeater's possum has an amazing history of rediscovery. Unlike the mountain pygmy-possum, which was once thought to have been extinct for thousands of years, Leadbeater's possum disappeared from view for just 52 years and was presumed lost forever. Leadbeater's possum is related to the smaller sugar glider, but it can be distinguished at once from that species because it lacks gliding membranes. It is grey to greyish brown above, with a prominent dark stripe along the back, and a paler colour underneath. The tail, which is wider at the tip than at the base, has an almost club-like appearance. The head and body length is 150-170 mm, with the tail adding a further 145-180 mm. Its weight range varies according to the season. In spring the range is 100-135 grams, and in the autumn 110-166 grams.

This possum appears to breed in all months except January and February. The most common months for births are May, June, October and November. The backwards-opening pouch of the female has four teats and litters of one or two are common. Female young leave the protection of the mother at about 10 months, and male young at about 15 months. This discrepancy may be explained by the fact that captive females have been observed to use force to disperse their own female offspring.

The diet of Leadbeater's possum consists of insects that live on foilage and under eucalypt bark, plant exudates (substances produced by trees at the sites of insect and other damage), and sweet secretions produced by sap-sucking insects. Variations in the animal's average weight can be explained by the greater abundance of food in summer. However, one important food source, a large tree-cricket, is available all year.

The secretive and shy nature of Leadbeater's possum, along with its dense, tall habitat, makes it.difficult to observe in the wild and partly accounts for the lack of sightings in the first half of the twentieth century. It is also fast-moving and, despite its lack of a gliding membrane, can often make impressive leaps of a metre or more from tree to tree. One aspect of this possum's behaviour is uncharacteristic of most mammals. It is the adult female, and not the males, that show most aggression socially. They will not allow unrelated females to enter and remain in their nests, even though males are permitted. As was mentioned above, captive females also evict female offspring at ten months, but do not molest male young. A communal nest is built in the hollow of a large mountain ash *(Eucalyptus regnans),*

Vulnerable	*Could become endangered because few in numbers or limited in distribution.*
Endangered	*Likely to become extinct unless action is taken.*
Critical	*Only a few hundred individuals survive, or only a single small population known.*
Extinct	*Not sighted for more than 50 years.*

LEADBEATER'S POSSUM

Gymnobelideus leadbeateri

Head and body length: 150-170 mm
Tail length: 145-180 mm

Leadbeater's possum does not have the gliding membrane found in the closely related sugar glider. Its tail is wider at the tip than at the base, giving it a club-like appearance.

(P. R. Brown/NPIAW)

Leadbeater's possum disappeared from view for 52 years and was presumed lost forever.

usually at 10-30 metres above the ground. The nest is lined with shredded bark and houses a monogamous breeding pair with one or more generations of offspring. The colony may also include unrelated males. The members of the colony engage in mutual grooming and seem to recognise each other by smell. They will protect their territory, which extends for one or two hectares, from others by attacking and biting.

The history of the enigmatic possum gives hope to those who seek the re-emergence of 'extinct' species. Only five specimens were collected between 1867, when it was first discovered, and 1909. These were collected from two sites in South Gippsland, Victoria — on Bass River and Koo-Wee-Rup Swamp — and from Mount Wills, 250 km away in northeastern Victoria. Since it was rediscovered in 1961 near Marysville, Victoria, it has been found at more than 50 localities scattered over 1000 km^2 of forest. All of these more recent discoveries are from an area that stretches about 60 km from Mount Torbreck to Mount Baw Baw in southeastern Victoria. It is not known why its populations are so scattered and why it is sometimes not found in adjacent areas that have the same type of forest. Within its distribution it is usually found in a relatively high density of 1.5-3 animals per hectare. Much of the animal's known range has been subject to fires and it seems to require the dense understory of regrowth in order to survive. However, it is now thought that the requirement most essential to its existence is the presence of large old or dead trees which provide nesting sites. As most forestry operations eliminate such trees, consideration needs to be given to maintaining areas with tree hollows in suitable mixed-age forest.

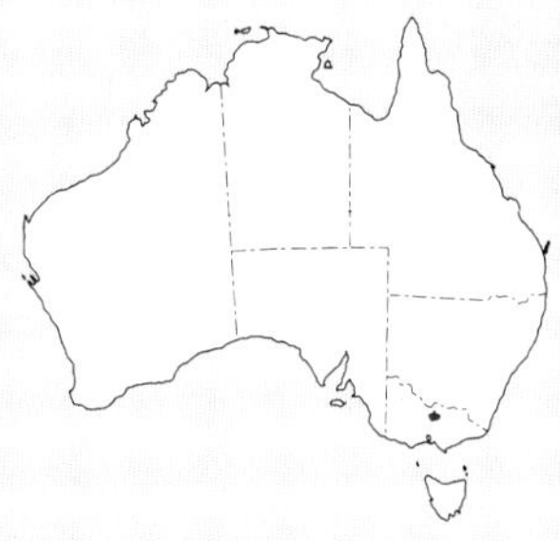

Victoria: An area that stretches about 60 km from Mount Torbreck to Mount Baw Baw.

Although the lack of recorded sightings or specimens of Leadbeater's possum for 52 years may be attributable in part to the animal's secretive nature, and to lack of searching, it appears that the possum has experienced a reduction of habitat. As its remaining habitat, apart from limited water catchment areas, is in a forestry concession designated for progressive clearfelling on a 40-80-year cycle, the future of Leadbeater's possum is anything but secure. There are no records of this possum recolonising a clear-felled and regenerating site. Its limited range and threats to the habitat of the species put Leadbeater's possum in the endangered category. Drastic measures, such as the setting up of reserves and the implementing of some changes in forestry practices, will need to be taken to ensure its survival.

LONG-FOOTED POTOROO

(Potorous longipes)

The long-footed potoroo inhabits rainforest gullies in areas of wet sclerophyll forest in eastern Victoria. The region is now under threat from woodpulp and other forest industries. It would be wise to protect most of the region in a reserve of some kind.

(M. Trenerry/NPIAW)

It digs conical pits in the soil to obtain its food, which probably includes roots, insects and other invertebrates.

It is not often that a new species of largish mammal is discovered in the late twentieth century. This makes the discovery of the long-footed potoroo, less than 15 years ago, a remarkable event. Until its discovery no new mammals had been found in Victoria since 1888. The first specimen to be recognised as a new species was caught in a dog trap near Bonang. Described scientifically by Seebeck and Johnston in 1980, this potoroo weighs approximately 1.6-2.2 kg and has a head and body length of 380-415 mm and a tail length of 315-325 mm. Males are more robust and are heavier than females. The general colour of the long-footed potoroo is greyish brown on the back and pale grey on the belly. The sides of the head are more greyish and there is a distinct facial stripe. The sparsely haired tail is dark brown above and pale brown below. The feet and hands are also sparsely haired and pale brown, and the front claws long and stout. The back feet are longer than in the other two species of potoroo, hence the animal's common and scientific names. As in other potoroos, pads are present at the bases of the toes. However, the long-footed potoroo has an extra pad. This species has 24 chromosomes in both sexes. In this respect it differs from the more common long-nosed potoroo, which has 12 chromosomes in the female and 13 in the male.

The female has a well-defined pouch with four nipples, although only one young has been recorded at a time. Early pouch young have been observed in June and October, and a lactating female was recorded in April. Babies were born in captivity in June, August and September. It seems probable that breeding takes place throughout the year. Although the diet of this animal in the wild has not been clearly determined, it readily eats fruit, sunflower seeds and commercial dog pellets in captivity. It digs conical pits in the soil to obtain its food, as does the long-nosed potoroo, and it seems probable that its diet, like that of the more common species, includes roots, insects and other invertebrates. However, there is a slight difference in the jaw articulation between the species and this may indicate some difference in diet.

Observations on the long-footed potoroo in the wild are limited and a great deal of what is known about the species has been collected from studies of a captive population. In captivity the long-footed potoroo will readily climb tree stumps, branches and artificial structures which it uses as vantage points from which to survey its surroundings. It descends in a more ungainly fashion. Despite its longer feet, its

Vulnerable	*Could become endangered because few in numbers or limited in distribution.*
Endangered	*Likely to become extinct unless action is taken.*
Critical	*Only a few hundred individuals survive, or only a single small population known.*
Extinct	*Not sighted for more than 50 years.*

LONG-FOOTED POTOROO

Potorous longipes

Head and body length: 380-415 mm

Tail length: 315-325 mm

Only a few photographs of this species have ever been taken. The long-footed potoroo's back feet are longer than those of the other two species of potoroo.

(Fredy Mercay/ANT Photo Library)

A 'kiss, kiss' sound may be heard between mother and young or when animals are under stress.

gait is similar to that of the long-nosed potoroo. A 'kiss, kiss' sound may be heard between mother and young or when animals are under stress. Both sexes show aggression towards each other when they are housed in small enclosures. Like other potoroo species, the animals will build nests from grass or hay that is provided for them.

The long-footed potoroo has so far been found as a living animal only in a small part of eastern Gippsland, Victoria. It may have eluded discovery for so long because of its extremely limited range, or because it was mistaken for the more common and widespread long-nosed potoroo, which also inhabits this area. Only ten specimens have been collected from a few localities. Two sites can be precisely located: Bellbird Creek, 32 km east of Orbost; and Yalmy Road, 26 km southwest of Bonang. The animal has also been recorded, just over the Victorian border, in the Bondi State Forest in New South Wales. This record was based only on some fur which, along with a jaw fragment containing two teeth, was found in fox droppings.

The long-footed potoroo lives in open forest of mixed eucalypts with a sparse to dense understory of sclerophyllous shrubs and a layer of wiregrass, ferns and sedges. The area occupied by this species has cool to moderate summers, cold winters and an average annual rainfall of 1100-1200 mm. Altitudes of the known sites range from 150 to 600 metres.

The exact status of the long-footed potoroo has yet to be determined. Its geographic range may in fact be greater than indicated at present and it is not known whether its numbers or ranges have diminished in historic times. Apart from the Bondi State Forest, none of the sites of collection is included in a national or state park or conservation area of any kind. In many parts of its range it is under threat from the woodpulp and other forest industries. Forest that adjoins recent collection sites may also be subject to forest industry activity. Until more is known the long-footed potoroo must be considered to be rare and highly endangered. It would be wise to include much of the animal's known range in a protected reserve of some kind.

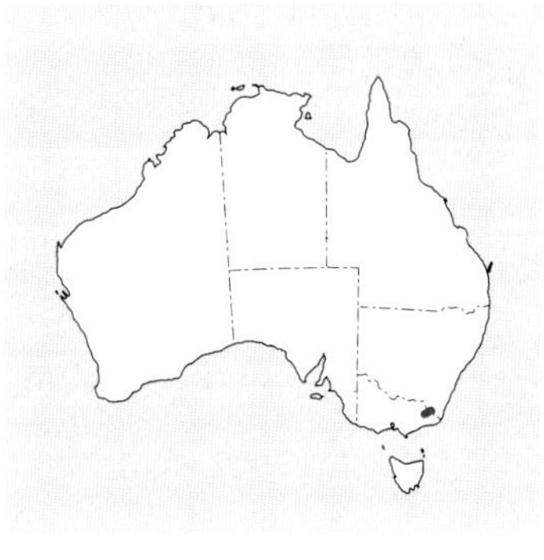

Victoria: A small part of eastern Gippsland.
New South Wales: Bondi State Forest.

LONG-TAILED DUNNART

(Sminthopsis longicaudata)

Habitat of the long-tailed dunnart. It is believed that this creature lives in rocky regions and, because it is seldom found in fossil deposits, it is thought to have always been a rare animal.

The yellow trap and tape tied to the tree are used by scientists in surveying rare animals such as this. A specimen was collected at Pillendinnie, near Marble Bar and donated to the West Australian Museum in 1940 and it was not until 1975 that another specimen was caught.

(B. Miller/NPIAW)

This dunnart is a good climber in its rugged, rocky habitat.

The long-tailed dunnart has proved an elusive and enigmatic creature. It is a most striking animal with a thin tail that is twice the length of its body. Its head and body length is 80-100 mm, while its tail is 180-210 mm long. Both males and females have a weight range of 15-20 grams. These measurements are based on an extremely limited number of specimens.

Long-tailed dunnarts are grey above and creamy white underneath. Their legs and feet are white and their exceptionally long tail is scaly with short hairs, except at the tip where long hairs protrude and make a fine brush. This dunnart also possesses striated foot pads which, along with its long tail, point to it being a good climber in its rugged, rocky habitat. This has proved to be the case in captive animals. The tail is also strongly muscular at its base and can be moved into a variety of positions. The long-tailed dunnart has a flattened head and a long snout. Its relatively large eyes and ears are common attributes of dunnarts that live in the arid zone.

Because only a small number of individuals have been located knowledge of the biology of this species is limited. Studies on captive animals show that both sexes can mate more than once in each breeding cycle, and both sexes maintain breeding condition from August to December. Females may come on heat up to four times within this period at intervals of roughly one month. Although the gestation time is not known precisely it is less than 20 days. The female has six teats in a pouch that fully encloses the young for the first three weeks of life. One litter of five has been recorded for a captive female. Indeed, the first specimen ever collected (in 1895) was an adult female with five pouch young. Sexual maturity may be reached from around eight to eleven months and individuals can live and breed for more than one year.

Analysis of the droppings of the long-tailed dunnart has shown that it eats a variety of beetles, ants, spiders, cockroaches, centipedes, grasshoppers, flies and various insect larvae. It may become torpid during cold periods.

The apparent prefered habitat of this species, which has been collected and recorded only in very few places, is rocky, arid areas. Before 1975 there were only three whole specimens in museums and nothing was known about its habits. Even details of where these three were captured were sketchy. The species was not scientifically described until 1908, when Professor Baldwin Spencer published upon

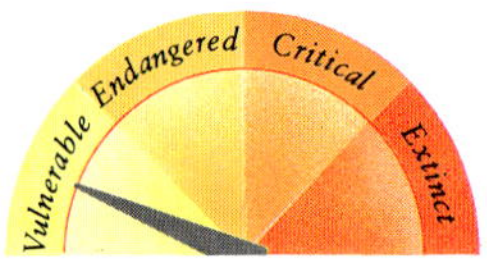

Vulnerable — *Could become endangered because few in numbers or limited in distribution.*

Endangered — *Likely to become extinct unless action is taken.*

Critical — *Only a few hundred individuals survive, or only a single small population known.*

Extinct — *Not sighted for more than 50 years.*

LONG-TAILED DUNNART

Sminthopsis longicaudata

Head and body length: 80-100 mm

Tail length: 180-210 mm

The long-tailed dunnart is found in central Australia. The function of its extraordinarily long tail is still not fully understood, but it may help in climbing.

(Babs & Bert Wells/NPIAW)

a specimen collected some time in 1896 in the Pilbara, Western Australia. Spencer already had a long-tailed dunnart in his own collection. It was a female with five pouch young that had been collected in central Australia in 1895, but it was not recognised as a long-tailed dunnart until 1916 after it was donated, along with other specimens, to the Museum of Victoria. A third specimen was collected at Pillendinnie, near Marble Bar, and donated to the Western Australian Museum in 1940. As well, some sub-fossil remains had been found in limestone caves in the Cape Range in Western Australia, and a piece of skullbone had been discovered near Marble Bar. A probable fossil has also been recorded from a cave in the MacDonnell Ranges, Northern Territory.

It may become torpid during cold periods.

Apart from these discoveries, no news was heard of the species until 1975. In that year the fourth adult specimen ever to be caught and recorded was taken from a spinifex hummock near Miss Gibson Hill in the Gibson Desert, Western Australia. The site was hummock grassland growing on sandy, gravelly soil with rocky breakaways nearby. Below a large breakaway ten metres from the capture sight grew low, open *Acacia* shrubs. Another six years were to pass before more specimens were located, in the Gibson Desert Nature Reserve. This discovery was the result of an expedition mounted by the Western Australian Wildlife Reserach Centre in order to locate further populations of this most elusive species. Nine animals were trapped — eight of them from plateaus, near breakaways and screes, and the other from a very rugged area of scree. The plateaus, composed of boulders, stones and red soil, were sparsely vegetated with mulga *(Acacia aneura)*, mini-ritchi *(Acacia grasbyi)* and spinifex. The scree consisted of a low open woodland of mulga.

Western Australia: May be widely distributed throughout rocky regions in the central western desert area.
Northern Territory: Uncertain distribution.

Today, it seems that the long-tailed dunnart may be widely distributed throughout rocky regions in the central western desert of Western Australia. Its distribution in the Northern Territory is uncertain and it is considered a rare species in both regions. It is seldom found in fossil deposits and may always have been rare. Although very few individuals have been recorded, not enough is known at this stage to say conclusively that the species is critically endangered.

Periods of hibernation may last up to a week at a time and are interspersed with bouts of feeding.

The discovery of a living mountain pygmy-possum was a most remarkable and unexpected event. This is because it was first described from fossil remains in 1895, and for over 70 years, until 1966, was thought to have become extinct along with the huge diprotodons many thousands of years ago.

This pretty little possum has two different colour phases, depending on its location. In both populations the fine, dense fur is grey-brown above and sometimes darker from the mid-back region to the top of the head. There is also a dark ring around the eye and a long, thin, scaly tail that appears naked but is clothed in sparse, fine short hairs. The Kosciusko population has a pale grey-brown or pale grey underside while in the Mount Hotham population it is more cream-coloured. The stout body has short limbs with prehensile fingers and toes. One of the most striking and distinguishing charcteristics of the mountain pygmy-possum for scientists is its enlarged and grooved posterior premolars. These teeth have a shearing action on food and are unlike the teeth of other pygmy-possums. However, a number of long-extinct mammals, including some contemporaries of dinosaurs, had similar teeth. The head and body length of this possum has been measured for two individuals as 115 mm for a male and 100 mm for a female, while an average has been given as 120 mm. The tail length has been measured as 148 mm for one male, 140 mm for one female, with an average recorded as 160 mm. The average body weight is 45 grams.

The mountain pygmy-possum has a rather curious sex life. For most of the year the males and females appear to live apart from each other. Adult females permanently inhabit the best locations on the rocky slopes while the males live on the margins. The areas occupied by the females can support a greater population and they have a better chance than the males of surviving more than one season. The dense vegetation in the female's habitat provides more food and protection. This breeding strategy sees the males 'migrating' to the female's habitat in order to breed and then returning to their habitat when the young are being weaned. This bizzare but successful method of reproduction has necessitated the building of rock-strewn 'possum underpasses' which allow males to join females in areas where the two habitats have been separated by roads.

Females have four teats and usually bear a litter of four. In the wild the young are born in November and December, although litters

Vulnerable — *Could become endangered because few in numbers or limited in distribution.*

Endangered — *Likely to become extinct unless action is taken.*

Critical — *Only a few hundred individuals survive, or only a single small population known.*

Extinct — *Not sighted for more than 50 years.*

MOUNTAIN PYGMY-POSSUM

(Burramys parvus)

During the winter the mountain pygmy-possum inhabits burrows below the snow in rocky areas such as this near Mount Kosciusko. The adult females inhabit the best locations on the rocky slopes while the males live on the margins of the female area. The dense vegetation in the habitat of the females provides more food and protection. The males only visit in order to breed. After the young are weaned, they return to their own habitat.

(J. & E. S. Baker/ANT Photo Library)

MOUNTAIN PYGMY-POSSUM

Burramys parvus

Head and body length: 100-115 mm
Tail length: 160 mm

The mountain pygmy-possum lives in boulder piles in Australia's snow country. Sometimes it has a darker patch of fur in the mid-back region.

(G. B. Baker/NPIAW)

Its ancestors used to live in rainforests, but the mountain pygmy-possum is now found in an alpine environment.

as early as August are known in captive populations. Young vacate the pouch when they are only a few weeks old and are thereafter left in a nest. At this time their eyes are still closed and their fur is short and sparse, but they grow rapidly and reach full size within five months. The nest is built by the female, mainly with thin-stemmed grasses from standing tussocks. After the grass is cut by the possum's specialised premolars, it is passed underneath and held in the curled-up tail until there is sufficient of it to carry, by means of the tail, to the nest. The mountain pygmy-possum is nocturnal and sleeps during the day within the nest, huddled in a ball shape with its head against its chest or abdomen and its tail curled up. Although the oestrous cycle and gestation periods are yet to be determined captive animals will breed in their second year. On Mount Higginbotham females have been known to breed in their first year, although some males seem to take at least two years to reach breeding condition.

The mountain pygmy-possum's diet consists of seeds, fruit and insects. Captive individuals are known to store seeds and fruit in heaps in their nests and under bark or soil. They use their highly manipulative forepaws as an aid to eating by holding all food in one or both paws. The mountain pygmy-possum is found in alpine areas which are interspersed with low shrubs and grasses. The shrub cover at lower altitudes is quite dense and there are some trees, usually snow gums. There are also groups of boulders, mostly from old glacial moraines. For three months of the year the area is covered with snow, and the animal may move through the spaces between boulders or under the shrubs that have prevented snow falling to the ground. An important survival factor in such an extreme environment is the animal's ability to enter a state of torpor when it becomes extremely cold. These periods of hibernation may last up to a week at a time and are interspersed with bouts of feeding.

One of the amazing aspects of the discovery of the seemingly extinct mountain pygmy-possum was that it was found in an alpine environment. Almost all of the fossils of related animals represent species that lived in rainforests. How did this rainforest relic end up maintaining an existence in such a different environment? One possible answer is that as the continent dried out, and gum trees replaced the rainforest, most of the tiny possums became extinct. However, some rainforest plants had found refuge around the margins of Australia's snow country, and it was these plants, growing on great boulder-strewn slopes in the Australian Alps, that provided just the conditions that the possum needed. The bushes and rocks kept the ground free of snow, while the blanket of snow itself insulated the possum from the worst effects of winter.

The mountain pygmy-possum was once known from widely distributed fossil material in New South Wales and Victoria. The discovery of the first live mountain pygmy-possum was made in 1966 in the Monash University ski lodge on Mount Hotham in Victoria. In most other buildings it may well have been destroyed as an unwanted rat-like pest or at best ignored. Fortunately the finders sought identifcation of the animal, whose discovery astounded the scientific world. Since its discovery in 1966 the animal has been located at Mount Kosciusko in New South Wales. Finds have also been reported at Falls Creek in Victoria. Its habitat area on Mount Hotham measures 20 by 10 km and on Mount Kosciusko 30 by 10 km. In the Kosiusko National Park it is found at altitudes above 1400 metres, but is found in lower altitudes in Victoria. Its habitat on the Kosiusko summit extends to 2230 metres, Australia's highest point. The different populations are completely separated by lowlands.

The habitat for the mountain pygmy-possum is coming under increasing pressure from development for recreational and tourist uses. The New South Wales population is protected within a national park, but the Victorian populations are not within national parks and the area supports a burgeoning ski-ing industry. Fortunately, most development to date has been sympathetic to this unique possum's needs. Tragically, however, the greenhouse effect may eventually bring about a further shrinkage of its habitat, and perhaps even the extinction of this species.

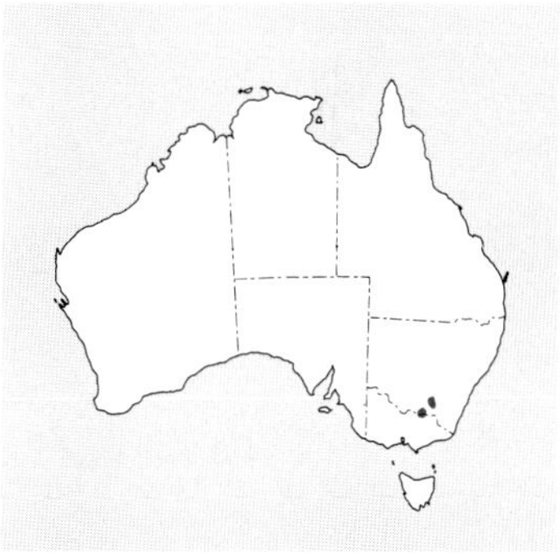

New South Wales: Mount Kosciusko, Kosciusko National Park.
Victoria: Mount Hotham.

NORTHERN HAIRY-NOSED WOMBAT

(Lasiorhinus krefftii)

Open woodland is the favoured habitat of the northern hairy-nosed wombat. Competition from cattle for grass badly effects this rare species. The surviving small colony of these animals has been incorporated into Epping Forest National Park.

(C. Andrew Henley/NPIAW)

It is known to renovate its burrow after rain.

The northern hairy-nosed wombat has had a very complicated scientific history. Its profusion of former common and scientific names, and frequent confusion between it and the other two wombat species, make its history since European settlement difficult to piece together. Although identified from fossil remains found at Wellington Caves in New South Wales in 1869, it was not widely recognised until Charles De Vis described some living animals in 1900. In fact De Vis thought that the Queensland specimens he described were a separate species from fossils and specimens that were collected in 1884 from the Deniliquin area in New South Wales. Many scientists now believe that they all belong to the same species. To complicate matters further, the scientist Professor F. W. Jones, writing in 1924, confused the scientific names of living and fossil species, causing the picture to become almost impossibly convoluted. It was not until 1937 that the only still existing colony was discovered at Epping Forest in mid-eastern Queensland, and subsequent study of these animals have clarified matters considerably.

The head and body length of the northern hairy-nosed wombat is approximately 1000 mm and its short, stumpy tail measures about 50 mm. Although no weights have been recorded, the animal is considerably larger than the southern hairy-nosed wombat which weighs 19-32 kg. It is largely browny-grey, or grey mottled with fawn, brown or black. Like the southern hairy-nosed wombat, its fur is softer and more silky, its nose more hairy and its ears longer than those of the common wombat. Its feet are well adapted to digging and its teeth grow continuously, as do those of all living wombat species. It differs markedly from the southern hairy-nosed wombat in many of its skull and facial characteristics. It has an extremely broad and oddly shaped nose with a robust, heavily built skull. In the central Queensland animals the bones that support the nose turn downwards at the tip.

Although some studies of the northern hairy-nosed wombat have been undertaken, little work has been published and details of its biology and behaviour remain sketchy. In common with the other wombats, it produces a single young at a time and has a backwards-opening pouch. This reduces the soiling of the pouch area that occurs when the wombat digs its burrow. The animal is nocturnal and spends its day in a burrow, emerging at night to feed. Diet includes grass and possibly other herbage. Its secretive nature has contributed to

Vulnerable	*Could become endangered because few in numbers or limited in distribution.*
Endangered	*Likely to become extinct unless action is taken.*
Critical	*Only a few hundred individuals survive, or only a single small population known.*
Extinct	*Not sighted for more than 50 years.*

NORTHERN HAIRY-NOSED WOMBAT

Lasiorhinus krefftii

Head and body length: approximately 1000 mm
Tail length: approximately 50 mm

This is one of the world's rarest animals. The northern hairy-nosed wombat is also one of Australia's largest surviving mammals. Being a nocturnal species, it spends the days in its burrow, emerging only at night to feed.

(C. Andrew Henley/NPIAW)

its elusiveness and goes some way to explain the comparatively recent discovery of its only surviving population. Dr Doug Crossman, who has studied the northern hairy-nosed wombat, has reported that it can spend up to six days within its burrow, which it is known to renovate after rain. Burrows occur both singly and in groups and their entrances are often strengthened by tree roots.

In historic times the northern hairy-nosed wombat is known to have occurred in three, or possibly four, widely scattered localities in the semi-arid interior of eastern Australia. No live animals have ever been reported in any intermediate areas. Fossil remains have been found in south-western and central New South Wales, as well as Victoria and Queensland. The favoured habitat appears to be flat, sandy, semi-arid grassland were there is eucalypt and acacia woodland as well as dense scrub and a ground cover of native grass. In 1884 four specimens were taken from a colony near Deniliquin in southern New South Wales. Until very recently, however, there had been no reports since 1909 and the species was presumed extinct in this area. Three specimens were collected about 1891 in southern Queensland, at two sites near St George, and there were reports of others near Tambo and Injune, in the same general area. However, after 1900 there were no further reports, and it was thought that the species may have died out entirely until, in 1937, a small colony was found at Epping Forest Station, in mid-eastern Queensland.

It has an extremely broad and oddly shaped nose with a robust, heavily built skull.

In 1985 some evidence appeared, suggesting that a wombat still existed in the former habitat of the northern hairy-nosed wombat near Deniliquin. Acting on local reports, the authors, accompanied by investigators from the Australian Museum and other institutions, conducted an expedition to investigate the wombat's existence. Two visits to the Tuppal State Forest, near Deniliquin, resulted in the location of about 25 burrows and burrow complexes. Although some of them appeared to be old — perhaps as old as 50 years — others were more recent. Fresh droppings and tracks were found, and hair taken taken from the droppings was identified as belonging to a hairy-nosed wombat. Unfortunately no further evidence has come to light, and it is possible that the hair belonged to a southern hairy-nosed wombat pet that had escaped from captivity. In the absence of further evidence, in the form of sightings, photos or specimens, the question of the continued existence of northern hairy-nosed wombats at Deniliquin remains unresolved.

The only known surviving population is restricted to its most northerly habitat, Epping and Waltham Stations, northwest of Clermont, in mid-eastern Queensland. This population, discovered in 1937, has recently been incorporated into Epping Forest National Park. The total area of the colony has been estimated at 15.5 km^2. Areas outside of the national park are subject to low intensity cattle grazing and there are no rabbits in the area. At the time the reserve was declared some 30-40 wombats inhabited the area. Recent surveys suggest that between 60 and 70 individuals now inhabit the reserve. Although the increase is encouraging, the northern hairy-nosed wombat still has perhaps the smallest population of any large mammal species on earth. All the increases in population size have involved reoccupation of old burrow systems within the same area. There is, therefore, no evidence that the species is extending its range.

The decline of the hairy-nosed wombat is the result, at least in part, of competition with cattle, sheep and rabbits. In the Deniliquin area many have been poisoned during a rabbit eradication program. It is to be hoped that the northern hairy-nosed wombat population can be protected from the effects of grazing, and that it will continue to grow and, perhaps even be reintroduced to parts of its once extensive range.

Queensland: Epping Forest National Park, northwest of Clermont.

NUMBAT

(Myrmecobius fasciatus)

FROM

The Mammals of Australia,

John Gould, 1848-1863

Gould's name for the mammal:

Banded Myrmecobius,

Myrmecobius fasciatus.

It has been observed on logs, tummy down and limbs and tail outstretched, soaking up the sun.

With its coarse, reddish brown coat and the white stripes on its blackish hindquarters, the numbat is one of the most distinctively coloured of Australian mammals. Although most closely related to the dasyurids (carnivorous marsupials), it is included in its own family, the Myrmecobiidae. The name numbat is derived from an Aboriginal word.

The numbat is about the size of a large rat, and weighs anything between 280 and 550 grams, depending on sex and age. Its head and body length has been measured as averaging 243 mm for males and 247 mm for females. Its tail adds another 160-210 mm for males and 160-190 mm for females. It has a flat head and rump and a long bristly tail which, when the hairs are erected, looks rather like a bottlebrush. Numbats have 50 or more teeth — more than any other marsupial — which, however, are very poorly developed and are used for eating only by the very young. The adults eat termites whole and break them up in the gut.

The female numbat does not have a pouch but has a mammary area with four teats surrounded by long hairs. Litters of two to four babies are born between January and March. Once the babies have detached from the teats they are left in the log or burrow, or ride on the mother's back. The young remain in their parents' territory for at least nine months.

The numbat lives almost exclusively on termites, although other invertebrates, such as ants, are also eaten, possibly by accident. Unlike other termite eating animals the numbat does not excavate the tough termite mounds but digs for termites in open spaces under leaf litter and around the bases of trees. Termite-infested logs are pulled apart with the sharp claws of the front feet. Underground termite galleries are located largely by scent and termites are eaten with a very long, pink, sticky, worm-like tongue. This flexible, cylindrical tongue is nearly half as long as its head and body and it gathers termites too fast for the human eye to see. Because a great deal of saliva is needed to keep the tongue moist and sticky the numbat has unusually large salivary glands. The fact that numbats are not able to utilise the vast food source of termites in mounds suggests that another species may have occupied this ecological niche. One author has suggested that a long-extinct species of echidna may have filled this ecological role.

The numbat is the only marsupial that is not at least partly

Vulnerable	*Could become endangered because few in numbers or limited in distribution.*
Endangered	*Likely to become extinct unless action is taken.*
Critical	*Only a few hundred individuals survive, or only a single small population known.*
Extinct	*Not sighted for more than 50 years.*

NUMBAT

Myrmecobius fasciatus

Head and body length: 243-247 mm
Tail length: 160-210 mm

Hollow logs provide numbats with refuges in times of danger, but the prevalence of bush fires threatens these numbat refuges. The animals are left open to attack from predators, such as foxes, which can dig them out of burrows (the alternatives when there are no hollow logs).

(C. Andrew Henley/NPIAW)

The numbat has a characteristic, jerky way of moving, and will often stop and remain motionless with its paw raised.

nocturnal. It is more active in the morning and late afternoon during the summer months and in the middle of the day during the autumn months. This diurnal behaviour pattern reflects termite behaviour patterns, for termites are more active in their feeding galleries by day than at night. Perhaps the colouration of the numbat is an adaptation to this daytime activity, in that it provides effective camouflage as protection against goannas and birds of prey, which are some of their natural predators. Since their introduction foxes, too, have become important predators.

Numbats usually shelter in fallen hollow logs or dead tree trunks and sometimes in burrows. Studies have shown that in some localities they may favour hollow logs during cold weather. They are certainly less susceptible to predation when using logs or trees because they have developed the ability to expand the chest and mid-body, and so wedge themselves inside the close fitting tree hollow. They can also turn inside the tightest of hollow logs, doing a kind of flip turn. The flattened nature of the rump aids in this manoeuvre. All this makes it very difficult for a predator to extract the animal. Nests of finely shredded bark have been found in logs and burrows, and it is likely that the otherwise obsolete teeth are used to shred this nesting material.

Today, the numbat inhabits woodland with grassy, herbaceous or shrub understory. Dominant trees in its habitat include the wandoo, *(Eucalyptus redunca)* and jarrah, *(Eucalyptus marginata)*. Mulga, *(Acacia aneura)* grows in areas where the numbat was formerly distributed, particularly in the more arid areas of southern Australia. Certain tree species are important as they provide food for termites; for the numbat feeds on the termites and uses hollow logs eaten out by termites for shelter.

The numbat is at least partly sociable as it lives in family groups for part of the year. Adults, however, forage alone. The numbat has some unique habits. It has a characteristic, jerky way of moving, and will often stop and remain motionless with its paw raised. Various authors have mentioned that when it is frightened the numbat has a curious habit of stamping the front feet one or two times, then raising and lowering their hindquarters and causing its tail hair to bristle before escaping very swiftly with its tail held horizontally — except for the last few centimetres where its is turned up. Many writers have noted its unaggresive, gentle behaviour when captured. It also appears to enjoy basking in the sun, and has been observed

Jarrah forest, such as this, and wandoo forest in the southwest of Western Australia provide the last refuge for the numbat.

(J. P. & E. S. Baker/ANT Photo Library)

on logs, tummy down and limbs and tail outstretched soaking up the warmth.

The numbat had a fairly extensive range when Europeans first settled the continent. It was found across a large part of southern Australia: in the western region of New South Wales around the Murray-Darling junction; in arid and semi-arid areas of South Australia, extending south to Adelaide; in the southern half of Western Australia; and even possibly, in the extreme northwest of Victoria and southwest of the Northern Territory. The great South Australian zoologist, Professor F. W. Jones proposed a separate South Australian subspecies in 1923, giving it the name *Myrmecobius fasciatus rufus*. Although the status of this subspecies is now often questioned, the central and eastern parts of the numbat's range did support individuals with a redder colouration, which was probably an adaptation to the red soil types common in such inland areas. At the time when Jones was writing, the South Australian population was most probably already extinct. Indeed, Jones predicted the extinction of the entire species: 'It will not be long before *Myrmecobius* will be as extinct as those Mesozoic Marsupials of the English Jurassic beds of which it has been said to be "actually an unmodified survivor".' Subsequent studies have shown Jones wrong in supposing the numbat to be a survivor of Mesozoic 'marsupials', and it is to be hoped that history

Many writers have noticed its unaggressive, gentle behaviour when captured.

will prove his prediction of extinction incorrect also. A former director of the Australian Museum, Gerard Krefft, noted the New South Wales numbat when he wrote about his 1856-57 expedition to the Murray-Darling area. Aborigines and station owners in the area did not consider the animal rare at that time, but this was the first and last time specimens were obtained from that area.

Termites are eaten with a very long, pink, sticky, worm-like tongue.

The present range of the numbat is restricted to the southwest of Western Australia. It is found in the Dryandra State Forest, the Perup Management Priority Area (east of Manjimup), and has only recently been discovered in several southern Perth suburbs. Knowledge of the populations in the Perth suburbs of Canning Vale, Forrestdale and Jandakot is based on sightings and road kills during 1985. One hopes that this area, which was allocated for housing at that time, still remains a viable habitat for the numbat.

One reason for the numbat's extinction in large parts of its former range could be that in many of these areas hollow logs were rare and the numbat was forced to make its home in burrows. This made it more susceptible to predation by introduced foxes, which could dig up the burrows. Other reasons suggested include the clearing of land for agricultural purposes and the cessation of Aboriginal made fire regimes. The smaller, controlled fires started by Aborigines reduced the incidence of larger, more devastating bushfires which could wipe out the numbat's habitat, food source and shelter.

A decline in sightings in the 1970s caused grave concern for the numbat's viability in its remaining refuges, but there seems to have been some recovery since then. It is, however, still considered an endangered species. Unfortunately, much of the numbat's present surviving habitat contains deposits of bauxite, and pressure to mine these may prove successful with future unsympathetic populations or governments.

Western Australia: Dryandra State Forest, the Perup Management Priority Area (east of Manjimup); Canning Vale, Forrestdale and Jandakot (southern Perth suburbs).

A captive breeding program was begun by the World Wildlife Fund Australia in the mid 1980s, in the hope of reintroducing individuals to protected habitats in their former range. Because of the difficulty of providing termites an artificial diet was being developed in 1985. Numbats have bred twice at Taronga Zoo, Sydney, where a colony was maintained for nine years.

One hopes that the selection of the numbat as the Western Australian faunal emblem does not mean it will suffer the same fate as the dodo and thylacine — the animals on the Mauritian and Tasmanian coat of arms respectively.

ORANGE HORSESHOE BAT

(Rhinonicteris aurantius)

FROM

The Mammals of Australia,

John Gould, 1848-1863.

Gould's name for the mammal:

Orange Horseshoe Bat,

Rhinolophus aurantius.

The lower part of the noseleaf forms a horseshoe shape below the nostrils.

This small, beautiful and usually orange bat was first described by Grey in 1845 from specimens collected at Port Essington, Coburg Peninsula, Northern Territory. There is a wide range of colour variation recorded in more recent studies. Although orange is common, individuals have been recorded with colours as diverse as dark reddish brown, dark lemon, pale lemon and even white. It has darker fur around the eyes. Its recorded weight is 8-10 grams, its head and body length is 45-53 mm, and its tail length 24-28 mm. As a further aid to identification, the forearm length is measured in scientific studies of bats. In the case of the orange horseshoe bat it is 47-50 mm.

The orange horseshoe bat is the only bat in its genus and is one of just two genera that are known to be unique to Australia. It is distinguished from other leaf-nosed or horseshoe bats by its rich colour and the peculiar scalloping of the upper margin of its noseleaf. The noseleaf is a striking facial feature which is quite complex, well developed and consists of two parts. The lower part forms the horseshoe shape below the nostrils and is made up of two leaves with a central gap at the front. The upper part of the noseleaf tops the nostrils like a crown and is extensively honeycombed and sculptured. This bat has short, pointed ears that cover the eye if laid forward and small, black eyes with pale eyelashes. The top of the head is raised well above the face. The wing membranes are a pale yellowish brown and limbs a pinkish yellow.

Little is known about the biology and behaviour of this bat, despite the urging of well-known naturalist authors throughout the twentieth century. Professor F. W. Jones, writing in 1924, and Ellis Troughton, writing in 1967, are two such scientists who saw the need for investigative projects. In spite of some studies that commenced in the 1980s, the animal's reproductive cycle still largely remains a mystery. The fact that it changes roosting caves on a regular seasonal basis may indicate a seasonal breeding cycle. Several specimens caught across the north of Australia from May to August 1981 revealed no juveniles, while pregnant or lactating females were also absent. This suggests that breeding may occur during the wet season in the months between October and April.

This bat is nocturnal, venturing out of its roost after sunset to feed on various invertebrates such as moths, beetles, shield-bugs, parasitic wasps, ants, chafers and weevils. It is known to return to its roost to digest its food, which suggests that most of its feeding

Vulnerable	*Could become endangered because few in numbers or limited in distribution.*
Endangered	*Likely to become extinct unless action is taken.*
Critical	*Only a few hundred individuals survive, or only a single small population known.*
Extinct	*Not sighted for more than 50 years.*

ORANGE HORSESHOE BAT

Rhinonicteris aurantius

Head and body length: 45-53 mm
Tail length: 24-28 mm

The orange horseshoe bat is variable in colour. Its beautiful fur can be fawn, brown, bright orange or golden, lemon, or even white, and it is soft and silky.

(G. D. Anderson/NPIAW)

It is known to return to its roost to digest its food, suggesting that most feeding is done fairly close to its home.

is done fairly close to its home. Although John Gould in one of the earliest references to the species (1863), stated that it roosts in the hollow trunks and spouts of eucalypts, all subsequent reports indicate a cave roost is usual. The animal hangs free or rests against the wall of its cave with its arms spread well apart. Orange horseshoe bats do not cluster for warmth as do some other bat species, but are usually found 12-15 cm. apart in colonies ranging from twenty to several thousand individuals. It will share caves with other bat species, particularly the dusky horseshoe bat *(Hipposideros ater)*. It has been shown to choose cave roosts with a hot and humid atmosphere — from 29-36°C and 95-100% relative humidity. One reason suggested for this preference is the bat's inability to maintain its body temperature when resting. In cooler roosts it would be susceptible to uncontrolled hypothermia and exhaustion. Compared to other bat species it is slow to rewarm and will become lethargic in temperatures below 20°C.

Colonies sometimes have to move, temporarily, from some roosts that are prone to flooding during the wet season. Exactly where they go remains a mystery. This factor may be responsible for Gould's report that the species roosted in hollow trees. There have also been reports that the species roosted temporarily in a household laundry and a shed.

This bat has been shown to travel faster than other types of horseshoe bats. Tagging of a number of orange horseshoe bats showed that they flew at an almost constant height of one metre above the spinifex; those studied in another location were observed to fly two metres above a thick scrub understorey and three metres above the ground surface. Like most other insectivorous bats, this species uses echolocation as an aid in locating food. Studies using a detector showed the call could be heard over a short range between one and two metres.

A wide variety of habitats in northern Australia.

This bat is known to range over a wide variety of habitats in northern Australia. In the Pilbara region of Western Australia it inhabits areas of spinifex grassland with *Eucalyptus camaldulensis* scattered along the creeks. In the Kimberley region of Western Australia its habitat is generally dry, open forest with *Melaleuca leucadendron* along the creeks; it also inhabits coastal mangroves in this region. At Tolmer Falls in the Northern Territory the habitat is dense palm forest with little understorey, surrounded by low open woodland. Grasslands with small scattered stands of low eucalypt woodland

The orange horseshoe bat inhabits caves formed in limestone in northern Australia. Limestone karst such as this at Cutta Cutta caves Nature Park near Katherine, Northern Territory is an important habitat.
(Denis & Theresa O'Byrne/ANT Photo Library)

is part of its habitat around the Kalkadoon Cave, Queensland. Professor F. W. Jones reported its existence in South Australia in 1924 and noted that it is was very rare. The explorer Edward John Eyre reported its existence in the centre of Australia. It is not certain if these reports resulted from confusion with another bat species or if the orange horseshoe bat's range once extended further south than it does now.

The orange horseshoe bat is considered to be an endangered species, in part perhaps because of our limited knowledge of its biology, behaviour and requirements. It appears to be particularly sensitive to disturbance and will completely abandon a roost which is subject to continual human interference. Even a slight disturbance will cause it to fly away or move further into the cave. Its range is threatened by development and tourism. Early attempts to develop the Cutta Cutta Cave in the Northern Territory for tourism led to a massive reduction in the bat population. In common with other tropical cave-dwelling bats they are prone to predation by snakes but this does not appear to be a threat to their overall survival.

RUFOUS HARE-WALLABY

(Lagorchestes hirsutus)

FROM

The Mammals of Australia,

John Gould, 1848-1863.

Gould's name for the mammal:

Rufous Hare-kangaroo,

Lagorchestes hirsutus.

The characteristic whistling sound made by the rufous hare-wallaby led early colonists to give it the name of 'whistler'.

The future of this rather shaggy, red-coated wallaby is very insecure. The rufous hare-wallaby has undergone a drastic shrinkage in range and its numbers are prone to fluctuation in its two remaining habitats. The scientific name *hirsutus,* meaning 'hairy', is a very apt description of this animal's long pelage. The head of the rufous hare-wallaby is usually dark grey, although in the Tanami Desert populations it is red. The tail is brownish black above and pale red below. The animal has a head and body length of 310-390 mm and a tail length of 245-280 mm. It weighs somewhere between 780 and 1960 grams with the females, on average, being larger and heavier than the males.

Information on the animal's reproduction is limited, although more is known about its diet. The rufous hare-wallaby appears to favour recently burned vegetation that is regrowing. The small population in the Tanami Desert eats seedheads and young sedge and grass leaves. Herbs and shrubs are also eaten. It has a stomach that is well adapted to a high plant-fibre diet. A short burrow, up to 70 cm long, is often dug for protection, and for shelter in hot weather. Alternatively, the animal may rest during the day, hidden under a hummock of spinifex.

John Gould's collector, John Gilbert, related this information about the animal's burrow and behaviour: 'It constructs a burrow, open at both ends, with a seat at the side of the entrance, from which it plunges into the burrow the instant it is alarmed. It feeds on the open country adjacent to the thickets, where there is low scrub about two feet high: when running, and particularly when hunted, it utters a singular note, resembling the syllable "ting" rather quickly repeated' (Gould 1863). This characteristic whistling sound made by the rufous hare-wallaby led early colonists to give it the name of 'whistler'. Its habit of hiding in spinifex has also given rise to its name of spinifex-rat. It is known to the Aborigines of the Tanami Desert, where it still survives, as the *mala.* At least one of its other Aboriginal names, *wurrup,* is now obsolete as a result of its local extinction.

The rufous hare-wallaby was once so common throughout most of the arid and semi-arid parts of Australia that many early explorers commented on its abundance. It was an important food source for Aborigines, being particularly common in spinifex hummock grasslands of the sand plain and sand dune deserts. There are records that it still existed in the Great Sandy and Gibson Deserts as recently as

Vulnerable	*Could become endangered because few in numbers or limited in distribution.*
Endangered	*Likely to become extinct unless action is taken.*
Critical	*Only a few hundred individuals survive, or only a single small population known.*
Extinct	*Not sighted for more than 50 years.*

RUFOUS HARE-WALLABY

Lagorchestes hirsutus

Head and body length: 310-390 mm

Tail length: 245-280 mm

The reddish colour of the coat is seen only in animals from central Australia — those from Western Australia are more grey.

When running, especially when hunted, the rufous hare-wallaby utters a single note resembling the syllable 'ting', which it repeats over and over again.

(L. F. Schick/NPIAW)

The rufous hare-wallaby leaves distinctive tracks in the bare ground between spinifex bushes. This photograph was taken in the Tanami Desert, Northern Territory. Today, the Tanami Desert Wildlife Sanctuary preserves this dry spinifex terrain. *(K. Johnson/NPIAW)*

the 1930s. Now its only remaining habitats are Bernier and Dorre Islands off Shark Bay in Western Australia and a small part of the Tanami Desert in the Northern Territory. The population in the Tanami Desert is localised around Salt Beef Lake in an area totalling about 100 km^2. The habitat in the Tanami Desert is characterised by low, deflated sand dunes and is dominated by the spinifex *Triodia pungens* and the tea-tree *Melaleuca glomerata*. These grow adjacent to salt lakes and drainage systems. Mounds of the termite *Nasutitermes triodiae* are also a feature of this landscape. Studies have shown that the animal is fairly mobile but is largely absent from areas of old spinifex, preferring areas that are regenerating after fires.

Numbers in all remaining habitats fluctuate significantly. A succession of good seasons before 1986 resulted in an increase in numbers, but its existence remains quite insecure, and its numbers have since declined in the Tanami Desert. Numbers on Dorre Island increased dramatically in an area which had been burned five years earlier. The disappearance of the rufous hare-wallaby appears to be directly linked to the cessation of Aboriginal fire regimes. Regulated fire is now being used in the Tanami Desert to encourage an increase in wallaby numbers and to prevent devastating bushfires that destroy the mature spinifex which is required for shelter. The cessation of Aboriginal fire regimes and the resulting wildfires also diminish the diversity and natural mosaic of the vegetation. It is to be hoped that, with the aid of good fire and wildlife management including a captive breeding program, new colonies of the rufous hare-wallaby will be able to be established and maintained.

RUFOUS HARE-WALLABY
Western Australia: Bernier and Dorre islands in Shark Bay.
Northern Territory: A small part of the Tanami Desert.

SHARK BAY MOUSE

(Pseudomys praeconis)

Mainland coast, Shark Bay. Only a single Shark Bay mouse was ever collected on the mainland, the others coming from Bernier Island.

(Denis & Theresa O'Byrne/ANT Photo Library)

It is remotely possible that the Shark Bay mouse is still living in some mainland areas.

Although very few specimens or sightings of this elusive creature were obtained before 1975, it now appears that the Shark Bay mouse may have been more common in its extremely restricted habitat than it was at first thought. Also known as the shaggy mouse, because of its particularly long, shaggy coat, the Shark Bay mouse has an unusual fur colouration. The underfur on the animal's back is a pale yellow-fawn while the dark guard hairs give the fur a grizzled dark brown appearance which grades into a delicate buff shade on the sides and into pure white below. The greyish ears are quite long and the large eyes are surrounded by a darkish ring of hairs. The feet are white, as is the underside of the fully furred tail. The top of the tail is grey and there is a tuft of dark hairs at the end. The colours of the tail are clearly delineated. The Shark Bay mouse has a head and body length of only 85-115 mm while its tail, which is longer than its body, is 115-125 mm. Its weight has been recorded as 30-50 grams.

Although there are no records of the breeding habits of the Shark Bay mouse in its natural state, observations of captive individuals give some limited information. Studies of a single female have shown that the oestrous cycle is 14 days or less. The animal gave birth to two litters — one of three and the other of four young — after gestation periods that varied from 28 to 30 days. The young clung determinedly to their mothers teats and were dragged around when she moved. This habit is common to many Australian rodents. Their eyes opened at 15 days and they were weaned by 30 days. The father behaved protectively towards the young by huddling over them on several occasions.

The diet of the Shark Bay mouse was determined by an examination of droppings found within traps that caught live specimens. It was found to consist mainly of the petals and antlers of a flower that is probably from the coast daisy bush *(Olearia axillaris)*. A small quantity of leaves and stems from an unknown fleshy plant was also found. In captivity the animal readily accepts a diet of mixed seeds and vegetables and tends towards obesity.

The available information on the habits of the Shark Bay mouse is very scant indeed. The animal appears to be fairly gentle; no aggression has been observed between members of the limited captive population. Recent data on captures in the wild seem to indicate that this is a fairly social species. Animals were trapped in areas

Vulnerable	*Could become endangered because few in numbers or limited in distribution.*
Endangered	*Likely to become extinct unless action is taken.*
Critical	*Only a few hundred individuals survive, or only a single small population known.*
Extinct	*Not sighted for more than 50 years.*

SHARK BAY MOUSE

Pseudomys praeconis

Head and body length: 85-115 mm

Tail length: 115-125 mm

This mouse is found only in one small area of coastal vegetation on Bernier Island, Shark Bay. It is also called the 'shaggy mouse' because of its very long, shaggy coat. The fur on its underside is pure white.

(H. & J. Beste/NPIAW)

where dense coastal spinfex *(Spinifex longifolius)* and the daisy *Olearia axillaris* grew on coastal sandhills. Some researchers think that the Shark Bay mouse may make its home in tunnels and runways in the vegetation, but this has not been confirmed as yet. When disturbed, the animal produces a faint whistling cry.

The very first known specimen, collected by Dr F. M. Rayner during the voyage of the *Herald* in 1858, was the only one to be located on the mainland. It was found on the Peron Peninsula, adjacent to Shark Bay in Western Australia. In 1906 a weathered skull was found on Bernier Island in Shark Bay. From then until 1975 less than six specimens were found, all of them on Bernier Island. During a 1975 rodent-collecting expedition, undertaken for the Institute of Medical and Veterinary Science (South Australia), 11 animals were located, ten of which were males, from Bernier Island.

Western Australia: Bernier Island in Shark Bay.

Fossil remains of the Shark Bay mouse have been found in cave deposits in a number of localities in Western Australia. These show that up to about 500 years ago the animal was fairly widely established on the mainland, and it is remotely possible that the Shark Bay mouse is still living in some mainland areas. The protection of Bernier Island, however, offers the best chance for the survival of the species. The island has been classified as a class 'A' fauna reserve with restricted public access. This is certainly a wise move in view of the number of vulnerable species that survive there. There have also been attempts to eradicate the feral goats that inhabit the island.

It is not known why the Shark Bay mouse has such a restricted distribution or why its range has shrunk since — as well, perhaps, as before — European colonisation. It could be that a number of factors, including climatic change and a change in fire regimes, have been responsible.

WESTERN BARRED BANDICOOT

(Perameles bougainville)

FROM
The Mammals of Australia,
John Gould, 1848-1863.
Gould's name for the mammal:
Saddle-backed Perameles,
Perameles myosurus.

Krefft commented that this mammal was as useful as a cat around the camp, being a proficient mouse catcher.

Vulnerable	*Could become endangered because few in numbers or limited in distribution.*
Endangered	*Likely to become extinct unless action is taken.*
Critical	*Only a few hundred individuals survive, or only a single small population known.*
Extinct	*Not sighted for more than 50 years.*

The western barred bandicoot has had various scientific and common names since first being described by the French naturalists Quoy and Gaimard in 1824. This may make it difficult to locate in some of the older accounts. Now known as *Perameles bougainville,* it has also been described under the names *Perameles fasciata, Perameles myosura* and *Perameles myosura notina.* These scientific names had been given to local populations and, because the various mainland forms are now extinct and few specimens were preserved, it is unlikely that their relationships and status will ever be fully understood. Furthermore, it has been known under as many as nine common names. One of those given in the past was erroneous and has led to some confusion. This was the name 'marl', which was derived from *mala,* a Western Australian Aboriginal name for the rufous hare-wallaby. Most of the limited information we have on the western barred bandicoot was collected during the nineteenth and the first quarter of the twentieth centuries. Few living zoologists are familiar with it.

The appearance of the western barred bandicoot varies throughout its former range, for both the colour and the distinctness of the bars on its back may differ between regions. It has a head and body length of 200-300 mm and a tail length of 190-250 mm. This lightly built, graceful and attractive bandicoot can weigh anything between 190 and 250 grams. It has two or three alternating light and dark bars on its rump. The Western Australian mainland form of this species had a dark patch on the rump and bars that were less distinct. The South Australian form had a blackish patch on the loins with three distinct black bars coming down from there to the thigh. The area between the bars is pale fawn. Both of the above forms of the animal were white to grey on the underside and feet. It has abundant underfur, especially towards the hind end of the body. The general colour of the body is a brownish-grey; the fur is long and has a spiny appearance. However, the zoologist Professor F. W. Jones reported that the coat was soft to the touch. The large ears are broad at the base and taper towards their oval tips. They are thin and membranous and are carried erect. The tail is dark above, finely tapered and covered with fine hairs.

The western barred bandicoot appears to be mainly nocturnal, usually emerging at dusk, although it has been seen occasionally in the day. Although no detailed studies of its diet have been made, it appears to be omnivorous, eating insects, seeds, roots, herbs and

WESTERN BARRED BANDICOOT

Perameles bougainville

Head and body length: 200-300 mm
Tail length: 190-250 mm

This is the western form of barred bandicoot. The bars across the rump are subtle in this subspecies, but these bars were more obvious in the now-extinct western barred bandicoots that came from further east.

(Babs & Bert Wells/NPIAW)

Western barred bandicoots were abundant on the saltbush covered Nullabor Plain as recently as the 1920s. Today they are extinct on mainland Australia.

(NATFOTO/ANT Photo Library)

small animals. A former director of the Australian Museum, Gerard Krefft, writing about his 1856-57 expedition to the lower Murray and Darling rivers area, comments that it was as useful as a cat around the camp, being a proficient mouse catcher. The individuals that he caught as live specimens were fed with various kinds of rodents which they killed with great speed: '. . . the *Perameles* would tumble the mice about with its fore paws, break their hind legs, and eat generally the head only. I have seen a single individual kill as many as twenty mice in a very short time, breaking their bones successively, after which it would begin to satisfy its hunger' (Krefft 1866).

Little is known about the breeding habits of the western barred bandicoot. It has a backwards-opening pouch, with four teats on island populations and eight on mainland populations. Jones reports that the nipples of mainland females are arranged in a circle which

was open in front. Two appears to be the most common litter size, although one and three have also been recorded. The South Australian populations were reported to breed from May to June, and island populations from April to June. There has been a report of a female dropping one young when alarmed. The jettisoning of young while fleeing is not uncommon among kangaroos but is rarely reported in other marsupials.

Jones has written extensively on some of the more unusual habits of the western barred bandicoot. He commented on the animal's intense activity and its ability to jump and rise vertically in the air. He considered it to be one of the most elusive of animals and noted its unique ability to disappear quickly when disturbed. Although western barred bandicoots are reputed to make gentle pets, Jones mentioned their extremely pugnacious nature with members of their own species. They fight by kicking and scratching with the long claws of their hind feet. On one occasion eight live specimens sent from Ooldea to Adelaide had fought to the death during the journey, losing most of their hair in the process! Among the corpses were found two male and two female pouch young. These were revived, but attempts at breeding proved disastrous. At first the sexes were kept separate, but one female killed the other, and a male bit off the survivor's tail. When it was placed with one of the males the surviving female had to be rescued — barely alive and minus her hair. After recovery she was placed with the other male, only to beat him to the punch by plucking all his hair out. He died two days later. Finally, this female bred freely with the first male but showed a tendency to kill and eat her babies even when they had grown to half adult size.

The western barred bandicoot formerly occurred over much of southern Australia in semi-arid and arid areas with a variety of vegetation types. These included plains and sand ridges with woodlands, saltbush, bluebush, desert acacia, shrublands and heath. In its one remaining habitat, on Bernier and Dorre Islands, it occurs in sandhills behind beaches. John Gould reported that it preferred dense scrub such as *Allocasuarina* thickets, in the southwest of Western Australia where it lived, sometimes in pairs, in well-camouflaged grass nests. The nest is made from grasses and other vegetation in a hollow dug under a low shrub. In New South Wales the species was reported to have inhabitated the Liverpool Plains and western areas through to the Murray-Darling Basin. Krefft reported its existence as far east

Although they make gentle pets, they fight with members of their own species.

Western Australia: Bernier and Dorre islands in Shark Bay.

Jones commented on its intense activity and its ability to jump and rise vertically in the air.

as the Sydney district. It is presumed extinct in these areas, the last specimens being taken from the Liverpool Plains in 1841, and the Murray-Darling river junction in 1857. It was known from north-western Victoria, in the Murray Valley region, but is presumed to be extinct there as well. In mainland Western Australia it was once common in the Perth district, in the wheat belt and north to Shark Bay. The last specimen recorded in mainland Western Australia was in 1909 at Onslow. In South Australia it occurred along the Murray River, near Adelaide, at the head of St Vincent's Gulf, and in the far west. The last specimens from anywhere in mainland Australia were taken at Ooldea in South Australia in 1922.

The western barred bandicoot is now restricted to Bernier and Dorre Islands in Shark Bay, Western Australia, where it is still common. The reasons for the decline of this species on the mainland are not totally clear. Vegetation throughout its previous mainland range has been affected by stock grazing, crop production and changed fire regimes. Throughout much of this area various alien animals have also been introduced. These include feral goats, rabbits and predators such as foxes, cats and feral pigs. The situation on their island habitats is quite different, which may account for its healthy state there. Sheep were never introduced to Dorre Island, and came to Bernier Island only for a short time at the beginning of the twentieth century. The situation with feral goats was similar: none was introduced to Dorre, and those on Bernier have been largely eradicated. Although feral cats were introduced on both islands, they are now reported as being extinct there. Given proper care, this species should remain viable on its island sanctuaries and should not suffer the fate of its mainland relations.

WESTERN QUOLL

(Dasyurus geoffroii)

FROM
The Mammals of Australia,
John Gould, 1848-1863.
Gould's name for the mammal:
Geoffroy's Dasyurus,
Dasyurus geoffroyi.

At birth they weigh about 15 mg and are no bigger than a grain of rice.

The western quoll was described scientifically by John Gould in 1841. However, it had long been known to Aborigines, and at least two Aboriginal names have been recorded for it. One of these, *kettrie,* was recorded in the nineteenth century by Gerard Krefft. The other, *chuditch,* sounds not unlike the aggresive call of the western quoll.

This animal can weigh anything between 700 and 1300 grams, depending on age and sex. It has a head and body length of 275-360 mm, and a tail length of 205-280 mm. The male is considerably larger than the female. Also known as the western native cat, this attractive animal is brown, with white spots on its upper body. It has a whitish underbody with thick, soft fur overall. No spots are present on its long crested tail. The ears are brown on the back, and white on the inside and on the outer edge. Generally, this species is very similar to the eastern quoll *(Dasyurus viverrinus).* The teeth of the two species are almost indistinguishable, but the western quoll has an extra toe on the hind foot and its tail is darker and less bushy than that of the eastern quoll.

The female western quoll has six nipples in a shallow pouch that begins to deepen about March in order to receive the young. In the following few weeks the inside of the pouch becomes stained with a reddish secretion, and the hairs become sparse and the nipples enlarge. The breeding season extends from late May to early July. The young are born after a gestation period of about 16 days. At birth they weigh about 15 mg and are no bigger than a grain of rice. Litter size is usually six but some individuals may be lost as they grow. By 11 weeks the babies are so big that they dangle from the pouch opening or cling to the mother's fur. By 15 weeks the young leave the pouch but still remain in the family group. They reach adult size and sexual maturity by 12 months.

The western quoll is nocturnal although it is most active around dawn and dusk. It can climb well and moves quickly on the ground. It usually spends the day in a tree hollow or a burrow dug by itself or another species. Burrows in sand dune country are dug as deep as 1.2 metres and have several entrances.

The western quoll's diet consists of insects and small vertebrates, including frogs, lizards, birds, hopping-mice and carrion. In settled areas it will raid poultry runs and rubbish bins, while in captivity it will eat meat, and even fruit, if it is offered. Regurgitation of food has been observed in captives; with vigorous abdominal pulsations

Vulnerable	*Could become endangered because few in numbers or limited in distribution.*
Endangered	*Likely to become extinct unless action is taken.*
Critical	*Only a few hundred individuals survive, or only a single small population known.*
Extinct	*Not sighted for more than 50 years.*

WESTERN QUOLL

Dasyurus geoffroii

Head and body length: 275-360 mm

Tail length: 205-280 mm

This animal appears to get enough fluid without drinking if it is on a diet of fresh meat.

It has a relatively low energy requirement, which should stand it in good stead in the arid parts of Australia.

(Babs & Bert Wells/NPIAW)

The western quoll saves energy by allowing its body temperature to drop when it is asleep.

the food goes from stomach to mouth where it is rechewed and reswallowed. During his 1856-57 expedition to the Murray-Darling junction, Krefft noted that the western quoll was an excellent vermin catcher. Referring to the mice and rats around camps and stations, he stated that '. . . there is no better remedy than to procure a few young Dasyuri . . . to effect a clearance . . .' (Krefft 1866).

The western quoll has a reputation for being quite ferocious. Krefft, however, may have exaggerated accounts of its aggression. He described the western quoll as fearless, bloodthirsty, solitary and a terror to birds. It would, he claimed, even eat its own progeny if nothing else was available. All his attempts to domesticate it proved fruitless, for after six months in captivity he found it to be more ferocious than ever! These observations have been contradicted to some extent by modern zoologists, who have found the species to be easy to keep and highly animated.

Some useful physiological adaptions to its environment have been recorded for the western quoll. The body temperature fluctuates from a daytime temperature of 34°C to a night temperature of 37°C, irrespective of external air temperatures. In fact, these body temperatures can be maintained and normal activity can continue even when the outside temperature is 0°C. Some individuals, however, react to cold mornings by becoming torpid for a short period. The western quoll thus saves energy by allowing its body temperature to drop when it is asleep. Body water is not lost readily, even if high environmental temperatures are encountered. It appears to get sufficient fluid without drinking if it is on a diet of fresh meat, and it has a relatively low energy requirement. These characteristics make it well suited to life in a wide range of climatic conditions.

Western Australia: Its only known habitat is in the southwest.

The original habitat of the western quoll was quite broad, ranging from desert to sclerophyll forest. Today it is found in eucalypt forest such as jarrah, and in the mallee, but it previously lived also in the sand dune region of arid central Australia. It was formerly widespread in South Australia from north to south and ranged as far east as Tarcowie in the Flinders Ranges. It is now thought to be extinct in South Australia. In the nineteenth century the western quoll was known from the centre, west and northwest of New South Wales, and the south central and southwestern parts of Queensland. It is now thought to be extinct there also. The same applies to the Northern Territory, although in the past it was known from Barrow Creek, Crown Point and Alice Springs, and was common to the south of

The forests of the southwest of Western Australia provide the last refuge of the western quoll.
(G. E. Schmida/ANT Photo Library)

Hooker Creek. It was also known south of a line from Shark Bay to Rawlinna in Western Australia. Today its only known habitat is in the south-west of Western Australia.

The western quoll's previous habitat had a wide diversity of botanical, geological and climatic attributes. For some time after 1979, when a *Dasyurus* species was discovered in New Guinea, the western quoll was thought to have lived there, but in 1988 the New Guinea specimens were found to represent a separate and quite distinct species.

It can climb well and moves quickly on the ground.

The reasons for the drastic decline of the western quoll throughout most of its range are unclear. It is, indeed, this drastic shrinkage of its range that has caused it to be considered an endangered species. Its rapid decline in central Australia coincided with the decline of the nomadic lifestyle of the Aborigines and the consequent cessation of the Aboriginal fire regime. This created conditions that were conducive to occasional severe and devasting wild fires. However, just why the western quoll should have declined in many areas but survived, like many other endangered animals, in the southwest corner of Western Australia is still not understood.

The list of threatened mammal species that follows is a compilation that updates the work of the Total Environment Centre's Endangered Species Group (*Our Wildlife in Peril*, Reed, Australia, 1983), and Ecofund Australia's subsequent review ('A National Threatened Species Strategy For Australia — Policies for the Future', 1986).

More recent information has been extracted from conservation reviews by the World Conservation Union's (IUCN) Species Survival Commission members in Australia (Species Action Plans), and other scientific institutions, government and non-government.

The system of classification used is based on the standard IUCN rating scheme for endangered and vulnerable species:

POTENTIALLY VULNERABLE: This category has been devised as a type of 'catch all' for species that would appear to need some form of conservation action or monitoring due to a 'general vulnerability' but which cannot be slotted into existing IUCN categories. This category is the one that provides us with the most challenging management problem for the decade ahead.

VULNERABLE: Taxa believed likely to move into the 'Endangered' category in the near future if the casual factors continue operating.

ENDANGERED: Taxa in danger of extinction and whose survival is unlikely if the casual factors continue operating.

PROBABLY EXTINCT: Not been reliably sighted in over 50 years.

World Wide Fund for Nature Australia (WWFA) is part of the world's largest private conservation organisations. WWF's mission is to achieve the conservation of nature and ecological processes by preserving genetic, species and ecosystem diversity, promoting the ecologically sustainable use of renewable natural resources, and eventually stop and reverse the degradation of our planet's natural environment.

Michael Kennedy
Senior Project Officer for WWF Australia

Mammals potentially threatened with extinction

PE *Probably Extinct*
E *Endangered*
V *Vulnerable*
PV *Potentially Vulnerable*

FAMILY NAME	SCIENTIFIC NAME	COMMON NAME	CONSERVATION STATUS
Thylacinidae	*Thylacinus cynocephalus*	Thylacine	PE
Dasyuridae	*Antechinomys laniger*	Kultarr	PV
Dasyuridae	*Antechinus godmani*	Atherton Antechinus	E
Dasyuridae	*Antechinus leo*	Cinnamon Antechinus	PV
Dasyuridae	*Dasyuroides byrnei*	Kowari	E
Dasyuridae	*Dasycercus cristicauda*	Mulgara	E
Dasyuridae	*Dasyurus geoffroii*	Chuditch (Western Quoll)	E
Dasyuridae	*Dasyurus maculatus*	Spotted-tailed Quoll	PV
Dasyuridae	*Dasyurus viverrinus*	Eastern Quoll	PV
Dasyuridae	*Parantechinus apicalis*	Dibbler	E
Dasyuridae	*Pseudantechinus rosamondae*	Little Red Antechinus	PV
Dasyuridae	*Phascogale calura*	Red-tailed Phascogale	E
Dasyuridae	*Phascogale tapoatafa*	Brush-tailed Phascogale	PV
Dasyuridae	*Sminthopsis aitkeni*	Kangaroo Island Dunnart	PV
Dasyuridae	*Sminthopsis butleri*	Carpentarian Dunnart	PV
Dasyuridae	*Sminthopsis longicaudata*	Long-tailed Dunnart	PV
Dasyuridae	*Sminthopsis psammophila*	Sandhill Dunnart	V
Dasyuridae	*Sminthopsis virginiae*	Red-cheeked Dunnart	PV
Myrmecobiidae	*Myrmecobius fasciatus*	Numbat	E
Peramelidae	*Chaeropus ecaudatus*	Pig-footed Bandicoot	PE
Peramelidae	*Echymipera rufescens*	Rufous Spiny Bandicoot	PV
Peramelidae	*Isoodon auratus*	Golden Bandicoot	E
Peramelidae	*Isoodon obesulus*	Southern Brown Bandicoot	PV
Peramelidae	*Perameles bougainville*	Western Barred Bandicoot	E
Peramelidae	*Perameles eremiana*	Desert Bandicoot	PE
Peramelidae	*Perameles gunnii*	Eastern Barred Bandicoot (mainland Aust.)	PV
Thylacomyidae	*Macrotis lagotis*	Bilby	E
Thylacomyidae	*Macrotis leucura*	Lesser Bilby	PE
Phalangeridae	*Trichosurus arnhemeneis*	Northern Brushtail Possum	PV
Phalangeridae	*Trichosurus caninus*	Mountain Brushtail Possum	PV
Burramyidae	*Burramys parvus*	Mountain Pygmy-possum	E
Petauridae	*Gymnobelideus leadbeateri*	Leadbeater's Possum	E
Petauridae	*Hemibelideus lemuroides*	Lemuroid Ringtail Possum	PV
Petauridae	*Petaurus australis*	Yellow-bellied Glider	PV
Petauridae	*Petauroides volans*	Greater Glider	PV
Petauridae	*Petaurus norfolcensis*	Squirrel Glider	PV
Petauridae	*Pseudocheirus archeri*	Green Ringtail Possum	PV
Petauridae	*Pseudocheirus herbertensis cinereus*	Daintree River Ringtail Possum	PV
Petauridae	*Pseudocheirus occidentalis*	Western Ringtail Possum	E
Macropodidae	*Dendrolagus bennettianus*	Bennett's Tree-kangaroo	PV
Macropodidae	*Dendrolagus lumholtzi*	Lumholtz's Tree-kangaroo	PV

FAMILY NAME	SCIENTIFIC NAME	COMMON NAME	CONSERVATION STATUS
Macropodidae	*Lagorchestes asomatus*	Central Hare-wallaby	PE
Macropodidae	*Lagorchestes conspicillatus*	Spectacled Hare-wallaby	PV
Macropodidae	*Lagorchestes hirsutus*	Rufous Hare-wallaby	E
Macropodidae	*Lagorchestes leporides*	Eastern Hare-Wallaby	PE
Macropodidae	*Lagorstrophus fasciatus*	Banded Hare-wallaby	E
Macropodidae	*Macropus eugenii*	Tammar Wallaby	V
Macropodidae	*Macropus greyi*	Toolache Wallaby	PE
Macropodidae	*Macropus irma*	Western Brush Wallaby	PV
Macropodidae	*Macropus parma*	Parma Wallaby	PV
Macropodidae	*Onychogalea fraenata*	Bridled Nail-tail Wallaby	E
Macropodidae	*Onychogalea lunata*	Crescent Nail-tail Wallaby	PE
Macropodidae	*Petrogale concinna*	Nabarlek	PV
Macropodidae	*Petrogale godmani*	Godman's Rock-wallaby	PV
Macropodidae	*Petrogale lateralis*	Black-footed Rock-wallaby	V
Macropodidae	*Petrogale penicillata*	Brush-tailed Rock-wallaby	V
Macropodidae	*Petrogale persephone*	Proserpine Rock-wallaby	E
Macropodidae	*Petrogale xanthopus*	Yellow-footed Rock-wallaby	PV
Macropodidae	*Setonix brachyrurus*	Quokka	PV
Potoroidae	*Bettongia gaimardi*	Tasmanian Bettong	PV
Potoroidae	*Bettongia lesueur*	Burrowing Bettong	E
Potoroidae	*Bettongia penicillata*	Brush-tailed Bettong	E
Potoroidae	*Bettongia tropica*	Northern Bettong	E
Potoroidae	*Caloprymnus campestris*	Desert Rat-kangaroo	PE
Potoroidae	*Hypsiprymnodon moschatus*	Musky Rat-kangaroo	PV
Potoroidae	*Potorous longipes*	Long-footed Potoroo	E
Potoroidae	*Potorous platyops*	Broad-faced Potoroo	PE
Potoroidae	*Potorous tridactylus*	Long-nosed Potoroo	PV
Phascolarctidae	*Phascolarctos cinereus*	Koala	PV
Vombatidae	*Lasiorhinus krefftii*	Northern Hairy-nosed Wombat	E
Vombatidae	*Lasiorhinus latifrons*	Southen Hairy-nosed Wombat	PV
Tarsipedidae	*Tarsipes rostratus*	Honey-possum	PV
Pteropodidae	*Dobsonia moluccense*	Bare-backed Fruit-bat	PV
Emballonuridae	*Taphozous australis*	North-eastern Sheathtail-bat	PV
Emballonuridae	*Taphozous mixtus*	Papuan Sheathtail-bat	PV
Megadermatidae	*Macroderma gigas*	Ghost Bat	V
Rhinolophidae	*Hipposideros cervinus*	Fawn Horsehoe-bat	PV
Rhinolophidae	*Hipposideros semoni*	Greater Wart-nosed Horsehoe-bat	PV
Rhinolophidae	*Hipposideros stenotis*	Lesser Wart-nosed Horsehoe-bat	PV
Rhinolophidae	*Rhinolophus philippinensis*	Large-eared Horseshoe-bat	PV
Rhinolophidae	*Rhinonicteris aurantius*	Orange Horseshoe-bat	V
Vespertilionidae	*Eptesicus douglasorum*	Yellow-tipped Eptesicus	PV
Vespertilionidae	*Eptesicus sagittula*	Large Forest Eptesicus	PV
Vespertilionidae	*Miniopterus australis*	Little Bent-wing Bat	PV
Vespertilionidae	*Miniopterus schreibersii*	Common Bent-wing Bat	PV
Vespertilionidae	*Murina florium*	Tube-nosed Insectivorous Bat	PV
Vespertilionidae	*Nycticeius influatas*	Hughenden Broad-nosed Bat	PV
Vespertilionidae	*Nictophilus walkeri*	Pygmy Long-eared Bat	PV

Dixon, J. M. 1988, 'Notes on the diet of three mammals presumed to be extinct: the Pig-footed Bandicoot, the Lesser Bilby and the Desert Rat Kangaroo', *Victorian Naturalist,* 105: 208-211.

Finlayson, H. H. 1955, *The Red Centre,* Angus & Robertson, Sydney.

Friend, J. A. & Kinnear, J. E. 1983, 'Numbat', in Strahan (ed.) op. cit., pp. 84-85.

Friend, T. 1982, 'The Numbat — An Endangered Species', *Australian Natural History,* Vol. 20, pp. 339-342.

Gibson, D. F. 1986, 'A Biological Survey of the Tanami Desertr in the Northern Territory', *Technical Report No. 30,* Conservation Commission of the Northern Territory, Alice Springs.

Gibson, D. F. 1986, 'The Tanami Desert Research on Aboriginal Land', *Australian Natural History,* Vol. 21, No. 6, pp. 544-546.

Gordon, G. 1983, 'Bridled Nail-tail Wallaby', in Strahan (ed.) op. cit., p. 205.

Gordon, G. 1983, 'Desert Bandicoot', in Strahan (ed.) op. cit., p. 102.

Gordon, G. 1983, 'Northern Hairy-nosed Wombat', in Strahan (ed.) op. cit., p. 122.

Gould, J. 1863, *The Mammals of Australia,* London (reissued 1983 incorporating the 3 original volumes published between 1845-63, with modern commentaries by J. M. Dixon, Macmillan, Melbourne).

Guiler, R. 1980, 'The Thylacine — An Unsolved Case', *Australian Natural History,* Vol. 20, pp. 69-70.

Guiller, R. 1985, *Thylacine. The Tragedy of the Tasmanian Tiger,* Oxford University Press, Melbourne.

Hickey, G. 1985, 'Numbats: Digging in for a Last Stand', *Australian Natural History,* Vol. 21, p. 272.

Johnson, K. A. 1983, 'Bilby', in Strahan (ed.) op. cit., pp. 107-108.

Jolly, S. 1983, 'Orange Horshoe Bat', in Strahan (ed.) op. cit., pp. 306-307.

Jones, W. 1923-35, *The Mammals of South Australia,* parts i-iii, Government Printer, Adelaide (reprinted 1968).

Krefft, G. 1866, 'On the Vertebrated Animals of the Lower Murray and Darling, Their Habits, Economy, and Geographical Distribution', *Transactions of the Philosophical Society of N.S.W. 1866:1-33.*

Mansergh, I. M. & Scotts, D. J. 1986, 'Winter Occurrence of the Mountain Pygmy-possum, *Burramys parvus* (Broom) (Marsupialia:Burramyidae), on Mt. Higginbotham, Victoria', *Australian Mammolgy,* Vol. 9, pp. 35-42.

Mitchell, T. L. 1839, *Three Expeditions into the Interior of Eastern Australia, etc.,* Vol. 2, T. W. Boone, London.

Nelson, J. 1989, 'Megadermatidae', in Walton & Richardson, (eds.) op. cit., pp. 852-855.

BIBLIOGRAPHY

Aitken, P. 1979, 'The Status of Endangered Australian Wombats, Bandicoots and the Marsupial Mole', in Tyler (ed.), op. cit., pp. 61-65.

Aitken, P. 1983, 'Pig-footed Bandicoot', in Strahan (ed.) op. cit., p. 104.

Archer, M. 1979, 'The Status of Australian Dasyurids, Thylacinids and Myrmecobiids', in Tyler (ed.) op. cit., pp. 29-43.

Archer, M., Flannery, T. & Grigg, G. 1985, *Kangaroo,* Weldon, Sydney.

Arnold, J. M. 1983, 'Western Quoll', in Strahan (ed.) op. cit. p. 22.

Aslin, H. J., Smith, M. J. & Ganf, R. W. 1987, *Marsupials of Australia,* vol. 2, *Carnivorous Marsupials and Bandicoots,* Lansdowne Editions, Sydney.

Bok, J. 1987, 'Rescuing the Stick-nest Rat', *Australian Geographic,* No. 6, pp.19-20.

Burbidge, A. A. 1983, 'Burrowing Bettong', in Strahan (ed.) op. cit., pp. 187-189.

Burbidge, A. A. 1983, 'Crescent Nail-tail Wallaby', in Strahan (ed.) op. cit., p. 206.

Burbidge, A. A. 1983, 'Western Barred Bandicoot', in Strahan (ed.) op. cit., p. 101.

Burbidge, A. A. & Johnson, K. A. 1983, 'Rufous Hare-wallaby', in Strahan (ed.) op. cit., p. 199.

Burbidge, A. A., McKenzie, N. L. & Fuller, P. J. 1983, 'Long-tailed Dunnart', in Strahan (ed.) op. cit., pp. 58-59.

Calaby, J. H. 1983, 'Mountain Pygmy-possum', in Strahan (ed.) op. cit., pp. 168-169.

Calaby, J. H. & Flannery, T. F. in press, *Marsupials of Australia,* Vol. 3, Lansdowne editions, Sydney.

Christensen, P. 1983, 'Brush-tailed Bettong', in Strahan (ed.) op. cit., pp. 184-185.

Churchill, S. K., Helman, P. M. & Hall, L. S. 1988, 'Distribution, Populations and Status of the Orange Horseshoe Bat', *Rhinonicterisaurantius,Australian Mammalogy,* Vol. II, pp. 27-34.

Dickman, C. 1986, 'Return of the Phantom Dibbler', *Australian Natural History,* Vol. 22, p. 33.

Dixon, J. M. 1983, 'Gould's Mouse', in Strahan (ed.) op. cit., p. 393.

Dixon, J. M. 1983, 'Long-tailed Hopping-mouse', in Strahan (ed.) op. cit., p. 435.

Dixon, J. M. 1983, 'White-footed Rabbit-rat', in Strahan (ed.) op. cit., p. 382.

FAMILY NAME	SCIENTIFIC NAME	COMMON NAME	CONSERVATION STATUS
Vespertilionidae	*Phoniscus papuensis*	Golden-tipped Bat	PV
Vespertilionidae	*Pipistrellus tasmaniensis*	Great Pipistrelle	PV
Muridae	*Conilurus albipes*	White-footed Rabbit-rat	PE
Muridae	*Leporillus apicalis*	Lesser Stick-nest Rat	PE
Muridae	*Leporillus conditor*	Greater Stick-nest Rat	E
Muridae	*Mastacomys fuscus*	Broad-toothed Rat	PV
Muridae	*Melomys cervinipes*	Fawn-footed Melomys	PV
Muridae	*Melomys hadrourus*	Thornton Peak Melomys	V
Muridae	*Mesembriomys gouldii*	Black-footed Tree-rat	PV
Muridae	*Notomys amplus*	Short-tailed Hopping-mouse	PE
Muridae	*Notomys aquilo*	Northern Hopping-mouse	V
Muridae	*Notomys fuscus*	Dusky-Hopping-mouse	V
Muridae	*Notomys longicaudatus*	Long-tailed Hopping-mouse	PE
Muridae	*Notomys macrotis*	Big-eared Hopping-mouse	PE
Muridae	*Notomys mitchelli*	Mitchell's Hopping-mouse	PV
Muridae	*Notomys mordax*	Darling Downs Hopping-mouse	PE
Muridae	*Pogonomys mollipilosus*	Prehensile-tailed rat	PV
Muridae	*Pseudomys albocinereus*	Ash-grey Mouse	PV
Muridae	*Pseudomys australis*	Plains Rat	PV
Muridae	*Pseudomys chapmani*	Pebble-mound Mouse	PV
Muridae	*Pseudomys desertor*	Desert Mouse	PV
Muridae	*Pseudomys fieldi*	Alice Springs Mouse	PE
Muridae	*Pseudomys fumeus*	Smoky Mouse	E
Muridae	*Pseudomys gouldii*	Gould's Mouse	PE
Muridae	*Pseudomys nanus*	Western Chestnut Mouse	PV
Muridae	*Pseudomys occidentalis*	Western Mouse	V
Muridae	*Pseudomys oralis*	Hastings River Mouse	V
Muridae	*Pseudomys pilligaensis*	Pilliga Mouse	V
Muridae	*Pseudomys praeconis*	Shark Bay Mouse	E
Muridae	*Pseudomys shortridgei*	Heath Rat	V
Muridae	*Rattus tunneyi*	Pale Field Rat	PV
Muridae	*Xeromys myoides*	False Water-rat	V
Muridae	*Zyzomys pedunculatus*	Central Rock-rat	E
Physeteridae	*Kogia breviceps*	Pygmy Sperm Whale	PV
Physeteridae	*Kogia simus*	Dwarf Sperm Whale	PV
Physeteridae	*Physeter macrocephalus*	Sperm Whale	E
Monodontidae	*Oracella brevirostris*	Irrawaddy Dolphin	PV
Delphinidae	*Sousa chinensis*	Indopacific Humpbacked Dolphin	PV
Balaenopteridae	*Balaenoptera acutorostrata*	Minke Whale	V
Balaenopteridae	*Balaenoptera borealis*	Sei Whale	E
Balaenopteridae	*Balaenoptera edeni*	Bryde's Whale	V
Balaenopteridae	*Balaenoptera musculus*	Blue Whale	E
Balaenopteridae	*Balaenoptera physalus*	Fin Whale	E
Balaenopteridae	*Megaptera novaeangliae*	Humpback Whale	E
Balaenidae	*Eubalaena australis*	Southern Right Whale	E
Ottariidae	*Neophoca cinerea*	Australian Sea-Lion	E
Dugongidae	*Dugong dugon*	Dugong	V

Colour separation by Rainbow Graphics Art Co. Ltd, Hong Kong
Printed and bound by Tien Wah Press (Pte) Ltd, Singapore

INDEX

Tyler, M. J. (ed.) 1979, *The Status of Endangered Australian Wildlife. Proceedings of the Centenary Symposium of the Royal Zoological Society of South Australia, Adelaide, 21-23, Sept. 1978.*

Walton, D. W. & Richardson, B. J. (eds.) 1989, *Fauna of Australia,* Vol. 1B, *Mammalia,* Australian Government Publishing Service, Canberra.

Watts, C. H. S.1983 'Dusky Hopping-mouse', in Strahan (ed.) op. cit., pp. 430-431.

Watts, C. H. S. & Aslin, H. J. 1981, *The Rodents of Australia,* Angus & Robertson, Sydney.

Winter, J. W. 1979, 'The Status of Endangered Australian Phalangeridae, Petauridae, Burramyidae, Tarsipedidae and the Koala', in Tyler (ed.) op. cit., pp. 45-60.

Woolley, P. A. 1983, 'Dibbler', in Strahan (ed.) op. cit., p. 29.

Newsome, A. E. 1962, 'Rabbit-eared Bandicoots or Bilbies', *Australian Natural History,* September, p. 97.

Ogilby, W. 1838, 'Proceedings of Meeting March 13th, 1838', *Proceedings of the Zoological Society of London, 1838,* pp. 25-27.

O'Rielly, H. M., Mansergh, I. & Willig, R. 1986, 'Daily Pattern of Activity of a Captive Mountain Pygmy-possum, *Burramys parvus* (Broom) (Marsupialia:Burramyidae), *Australian Mammalogy,* Vol. 9, pp. 53-55.

Pocock, R. L. 1926, 'The External Characteristics of *Thylacinus, Sarcophilus,* and Some Related Marsupials', *Proceedings of the Zoological Society of London, 1926,* pp. 1037-1084.

Poole, W. E. 1979, 'Australian Kangaroos', in Tyler (ed.) op. cit., pp. 13-27.

Prince, R. I. T. 1983, 'Banded Hare-wallaby', in Strahan (ed.) op. cit., pp. 201-202.

Richards, G. 1983, 'Ghost Bat', in Strahan (ed.) op. cit., pp. 292-293.

Robinson, A. C. 1983, 'Greater Stick-nest Rat', in Strahan (ed.) op. cit., p. 424.

Robinson, A. C. 1983, 'Shark Bay Mouse', in Strahan (ed.) op. cit., p. 392.

Robinson, A. C. & Young, M. C. 1983, *The Toolache Wallaby,* N.P.W.S. Special Publication 2, Department of Environment and Planning, South Australia.

Rounsevell, D. E. 1983, 'Thylacine', in Strahan (ed.) op. cit., pp. 82-83.

Seebeck, J. H. 1983, 'Long-footed Potoroo', in Strahan (ed.) op. cit., p. 183.

Smith, A. 1983, 'Leadbeater's Possum', in Strahan (ed.) op. cit., pp. 142-143.

Smith, M. J. 1983, 'Desert Rat-kangaroo', in Strahan (ed.) op. cit., p. 192.

Smith, M. J. 1983, 'Toolache Wallaby', in Strahan (ed.) op. cit., p. 234.

Smith, S. J. 1981, *The Tasmanian Tiger — 1980,* Report on an investigation of the current status of the Thylacine, *(Thylacine cynocephalus)* funded by the World Wildlife Fund Australia, Tasmanian National Parks and Wildlife Service, Report No. 81/1.

Spencer, W. B. 1987, 'Description of Two New Species of Marsupials from Central Australia', *Proceedings of the Royal Society of Victoria,* (new series) Vol. 9, pp. 6-11.

Strahan, R. (ed.) 1983, *The Australian Museum Complete Book of Australian Mammals,* Angus & Robertson, Sydney.

Strahan, R. 1983, 'Eastern Hare-wallaby', in Strahan (ed.) op. cit., p. 196.

Sturt, C. 1849, *Narrative of an Expedition into Central Australia performed under the authority of Her Majesty's Government during the Years 1844, 1845 and 1846,* T. & W. Boone, London.

Troughton, E. le G. 1941, *Furred Animals of Australia,* Angus & Robertson, Sydney.